# STRUCTURE, DYNAMICS, AND BIOGENESIS OF BIOMEMBRANES

# STRUCTURE, DYNAMICS, AND BIOGENESIS OF BIOMEMBRANES

Edited by
RYO SATO
SHUN-ICHI OHNISHI

JAPAN SCIENTIFIC SOCIETIES PRESS
*Tokyo*
PLENUM PRESS
*New York London*

CHEMISTRY

6825-8264

Published jointly by
JAPAN SCIENTIFIC SOCIETIES PRESS, Tokyo
ISBN 4-7622-5333-2
        and
PLENUM PRESS, New York  London
ISBN 0-306-41283-7
Library of Congress Catalogue Card No. 82-051051

Distributed in all areas outside Japan and Asia between Pakistan and Korea by PLENUM PRESS, New York  London.

Printed in Japan

# Preface

Biomembranes are not only principal constituents of the cell but also a major site of biological activity. Studies on biomembranes are therefore, crucial for furthering our understanding of life processes. A special research project on biomembranes involving more than 100 investigators was conducted under a grant from the Ministry of Education, Science and Culture of Japan between 1978 and 1981. This has resulted in marked progress on basic biomedical studies on biomembranes in Japan. This book is a compilation of several major developments made during this project in the field of "Structure, Dynamics, and Biogenesis of Biomembranes." Developments in other fields constitute the contents of a companion volume, "Transport and Bioenergetics in Biomembranes," edited by Sato and Kagawa. In these two volumes the authors review recent advances which have been primarily made in their own laboratories and include relevant work carried out by other investigators.

Eight topics are presented in this volume. Ikegami and coworkers review molecular dynamics in lipid bilayers, reconstituted membranes and biological membranes as studied by nanosecond fluorescence spectroscopy with an emphasis on the wobbling in the cone model. The following two chapters deal with virus-induced membrane fusions. Ohnishi and Maeda describe an assay method for envelope fusion using spin-labeled phospholipids and report several characteristic results obtained by this method including low pH-induced fusion of influenza virus. Asano and Asano analyze the virus-induced cell fusion process based on five elementary steps and discuss the mechanism of membrane fusions with an emphasis on the role of clustering of glycoproteins.

The regulation of surface receptor mobility by cytoplasmic microtubule and microfilament systems is reviewed by Yahara, who also discusses ligand-independent capping phenomena in lymphocytes induced in a hypertonic medium. Kobata summarizes recent developments in chemical structure determination, classification, and biosynthesis of sugar chains of glycoproteins and discusses their roles in cellular recognition. Mizushima gives a full account of reconstitution of the outer membrane of gram-negative bacteria and presents results of a study on the mechanism of bacteriophage infection.

The last two chapters are concerned with synthesis and translocation of cellular proteins. Kikuchi and Hayashi describe synthesis of mitochondrial matrix proteins, especially $\delta$-aminolevulinate synthase, in the cytoplasm and their transfer into the matrix across both the outer and inner mitochondrial membranes. Omura reviews the turnover, biosynthesis, and integration of microsomal membrane proteins and discusses the origin of the sidedness of the endoplasmic reticulum membranes.

All these topics may seem to be unrelated but are undoubtedly interwoven into a two-dimensional matrix under certain general principles. Molecular recognition and dynamic mobility of biomolecules and their control appear to be important common features for the functioning of biomembranes. We hope that this book, together with the companion volume, will provide an impetus to future developments in biomembrane research.

July 1982

Ryo SATO
Shun-ichi OHNISHI

# Contents

# Structure and Dynamics of Biological Membranes Studied by Nanosecond Fluorescence Spectroscopy

AKIRA IKEGAMI, KAZUHIKO KINOSITA, Jr.,
TSUTOMU KOUYAMA,[*1] AND SUGURU KAWATO[*2]

*The Institute of Physical and Chemical Research, Saitama 351, Japan*

Fluorescence spectroscopy is a useful technique which gives various information on the structure and function of biological membranes. Its applicability has been greatly enhanced in recent years by improvements in the technique for nanosecond fluorescence spectroscopy (*1*).

There are three fundamental applications of the nanosecond fluorescence spectroscopy in the study of biological membranes. First, fluorescence anisotropy decay is used to determine the rotational motion of molecules in membranes. Time-resolved measurements of fluorescence anisotropy decays have several advantages. Both dynamic and structural information can be extracted from the data in a straightforward way (*2*). Fluorescence depolarization measurements of hydrophobic probes embedded in lipid bilayers have been made to investigate the dynamic properties of lipid hydrocarbon chains and the nature of the phase transition (*3–7*), the effect of cholesterol on these properties (*6–9*), and the dynamic structure of various biological membranes (*10, 11*).

[*1] Present address: Cardiovascular Research Institute, University of California, San Francisco, California 94143, U.S.A.

[*2] Present address: Research Development Corporation, c/o Maruzen Sekiyu Research Center, Satte, Kitakatsushika-gun, Saitama 340-01, Japan.

1

As the second application, the dependence of the fluorescence lifetime and spectrum of a probe on its environment is used to detect the conformational changes in membrane. A clear dependence of the fluorescence lifetimes on the phase transition was observed in some probes (7, 12).

Since the rate of electronic excitation energy transfer between donor and acceptor chromophores depends strongly on their distance from each other, the technique was used to study the interactions between molecules in membranes. In particular, associations of peptides in membrane proteins (13–15) and the lateral distribution of lipids (16) and proteins (17) have been investigated using this technique. Intramembrane positions of fluorescent probes in several membranes (18, 19) were also estimated with this technique.

This article reviews our recent results obtained using nanosecond fluorescence spectroscopy to investigate lipid bilayers, reconstituted membranes, and biological membranes.

## I. WOBBLING MOTION OF THE HYDROCARBON CHAINS IN LIPID BILAYERS

We have studied the molecular motion of the hydrocarbon chains in various model and biological membranes with the fluorescence depolarization technique using 1,6-diphenyl-1,3,5-hexatriene (DPH) as a probe. The hydrophobic fluorescent probe, DPH, has a rod-like, all-*trans* polyene structure, and its absorption and fluorescence transition moments lie along the major axis of the probe. Therefore, DPH has been established as a probe for the molecular motion of hydrocarbon chains in membranes.

The nanosecond time-resolved fluorescence measurements were made by a multipath single photon counting fluorometer. Details were presented elsewhere (20).

*1. Nanosecond Fluorescence Depolarization and Wobbling in Cone Model*
*1) Fluorescence depolarization*
Fluorescence depolarization techniques can be used to measure the rotational motion of probes in membranes.

Rotational motion is the change in orientation of a probe with time. Excitation of probes with a vertically polarized light pulse produces an ensemble of excited probes in which the transition moments are preferentially aligned along the vertical direction. The probes then undergo Brownian motion and their orientations become randomized. If one observes the vertical $(I_V)$ and horizontal $(I_H)$ components of the fluorescense intensity, their time courses are complex functions of time and are dependent on both the lifetime $\tau$ and the rotational motion of the probe.

To separate the dependence on $\tau$ from that on rotational motion, the observed intensities are usually analyzed using the fluorescence intensity $I_T(t)$ and the fluorescence anisotropy $r(t)$ defined by:

$$I_T(t) = I_V(t) + 2I_H(t) \tag{1}$$

$$r(t) = (I_V(t) - I_H(t))/I_T(t) \tag{2}$$

The fundamental anisotropy $r_0$, the value of $r$ in the absence of rotation, is expressed by:

$$r_0 = 0.4 \left[ \frac{3 \cos^2 \lambda - 1}{2} \right] \tag{3}$$

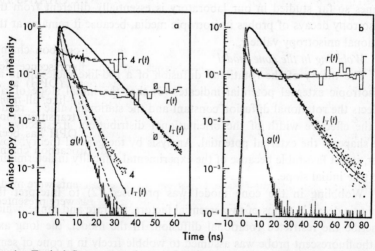

Fig. 1. Typical fluorescence decays of DPH in model and biological membranes. a) In pure dimyristoyl phosphatidyl choline (DMPC) vesicles (curves 0) and in vesicles of the oxidase-to-DMPC molar ratio of 4/100 (curves 4). b) Mitochondrial membranes. $g(t)$, the instrumental response function; $I_T(t)$, the total fluorescence intensity (dots, experimental data; dashed and solid lines, calculated best-fit curves for single- and double-exponential approximations); $r(t)$, fluorescence anisotropy (zigzag lines, experimental; smooth lines, calculated best-fit curves for the exponential-plus-constant approximation).

where $\lambda$ is the angle between the absorption and emission transition moments. The factor 0.4 is given for usual experimental conditions where the initial distribution of the chromophore is isotropic.

2) *Fluorescence anisotropy decays in membrane*

Typical results of the fluorescence decays of DPH embedded in model and biological membranes are shown in Fig. 1. The anisotropy decay of DPH, $r(t)$, is biphasic, consisting of an initial decrease and a following stational phase. The process can be expressed by the approximate form:

$$r(t)=(r_0-r_\infty)\exp(-t/\phi)+r_\infty \qquad (4)$$

Strictly speaking, many relaxation times should be included in the anisotropy decay from $r_0$ to $r_\infty$, and the first term in Eq. (4) should be replaced by the sum of several exponential terms. Unfortunately, we do not have enough experimental accuracy to separate many relaxation times.

The biphasic decay of anisotropy which was observed for all membranes so far studied in our laboratory is essentially different from the anisotropy decays of probes in isotropic media, because it remains at the stational anisotropy value $r_\infty$.

3) *Wobbling in the cone model*

The theory of the rotational diffusion of a rod-like molecule in an anisotropic external potential indicates (2) that the initial slope of $r(t)$ reflects the rotational diffusion constant and the stational value $r_\infty$ relates to the effective width of the orientational distribution, irrespective of the shape of the external potential. Analysis by the general theory, however, is not favorable because of the experimental difficulty in determining the true initial slope.

Wobbling in the cone model was proposed (2) to interpret the biphasic decay in terms of two essential factors, the rate and range, which describe the restricted rotational diffusion. In the model, the long axis of the fluorescent probe was assumed to wobble freely in a cone of semi-angle $\theta_c$ around the normal of the membrane with a wobbling diffusion constant $D_w$. Even in this simple model, $r(t)$ is expressed as the sum of an infinite number of exponentials, of which one is a constant, $r_\infty$. The calculated $r(t)$, however, can be replaced by a much simpler form:

$$r(t)/r_0=(1-A_\infty)\exp(-D_w t/\sigma_s)+A_\infty \qquad (5)$$

TABLE I.   Comparison between strict cone and Gaussian models (21).

| $r\infty/r_0$ | Strict cone model | | Gaussian model | | $\sigma_G/\sigma_s$ |
|---|---|---|---|---|---|
| | $\theta_c$ | $\sigma_s$ | $\theta_e$ | $\sigma_G$ | |
| 1.000 | 0.0 | 0.0 | 0.0 | 0.0 | |
| 0.989 | 5.0 | 0.0022 | 5.0 | | |
| 0.955 | 10.0 | 0.0088 | 10.0 | | |
| 0.901 | 15.0 | 0.0196 | 15.0 | | |
| 0.831 | 20.0 | 0.0342 | 20.1 | | |
| 0.746 | 25.0 | 0.0522 | 25.2 | | |
| 0.653 | 30.0 | 0.0731 | 30.3 | | |
| 0.555 | 35.0 | 0.0962 | 35.7 | | |
| 0.458 | 40.0 | 0.121 | 41.6 | 0.101 | 0.83 |
| 0.364 | 45.0 | 0.146 | 48.1 | 0.119 | 0.82 |
| 0.279 | 50.0 | 0.170 | 55.1 | 0.132 | 0.78 |
| 0.204 | 55.0 | 0.193 | 61.9 | 0.143 | 0.74 |
| 0.141 | 60.0 | 0.214 | 67.7 | 0.152 | 0.71 |
| 0.0904 | 65.0 | 0.231 | 72.3 | 0.158 | 0.68 |
| 0.0527 | 70.0 | 0.245 | 75.7 | 0.162 | 0.66 |
| 0.0265 | 75.0 | 0.253 | 78.1 | 0.165 | 0.65 |
| 0.0104 | 80.0 | 0.257 | 79.9 | 0.166 | 0.65 |
| 0.0022 | 85.0 | 0.255 | 81.2 | 0.167 | 0.65 |
| 0.0 | 90.0 | 0.250 | 82.2 | 0.167 | 0.67 |

where $A_\infty$ and $\sigma_s$ depend solely on $\theta_c$. Thus, if the experimental $r(t)$ is approximated by Eq. (4), the cone angle $\theta_c$ and the wobbling diffusion constant $D_w$ can be easily determined using Table I.

*4)   Gaussian cone model*

The wobbling in the cone model is simple, and can therefore be applied widely for the analysis of experimental results. To clarify its wide applicability, we examined the effect of the simple approximation of the model, the square well-type distribution of the probe orientation, on the analysis (21). A Gaussian distribution of the probe orientation was calculated and compared with the original cone model (see Table I). The results suggest that, when only two parameters are extracted from an experiment, the choice of a model is not crucial as long as the model contains the two essential factors describing the rate and angular range of the probe motion.

## 2.   Phase Transition of Lipid Hydrocarbon Chains

The temperature dependence of the steady-state fluorescence aniso-

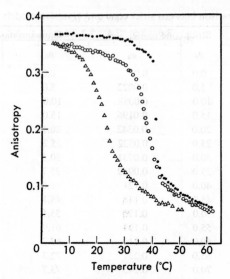

Fig. 2. Temperature dependence of the steady-state fluorescence anisotropy ($r^s$) of DPH in lecithin membranes. ● DPPC multibilayer liposomes; ○ DPPC sonicated vesicles; △ DMPC sonicated vesicles.

tropy $r^s$ of DPH embedded in the hydrocarbon regions of sonicated and unsonicated vesicles of saturated phospholipids, dimyristoyl phosphatidyl choline (DMPC) and dipalmitoyl phosphatidyl choline (DPPC), is shown in Fig. 2. Sharp changes in $r^s$ observed at about 40°C (DPPC) and 23°C (DMPC) correspond to the order-disorder phase transition. The corresponding changes in wobbling parameters, $\theta_c$ and $D_w$, were determined (see Fig. 3).

The obvious interpretation of the result is that the orientational distribution of DPH in hydrocarbon regions is highly anisotropic in the ordered crystalline state, and is anisotropic to a small extent even in the disordered state. The estimated values of $\theta_c$ are about 20° in the ordered state, sufficiently below the transition temperature $T_t$, and are about 70° above this temperature. These values indicate that the fairly large space must be attributed to the cone for the wobbling diffusion. It is very unnatural to consider that such a large space leaves a vacuum around each DPH molecule. A major part of the space should be occupied by hydrocarbon chains to reduce void volume. The probe wobbles around due to collision with these chains. The cone-type potential for wobbling

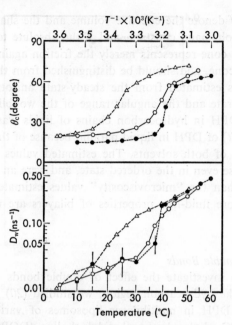

Fig. 3.  The wobbling diffusion constant, $D_w$, and the cone angle, $\theta_c$, for the wobbling motion of DPH in lecithin membranes. ● DPPC multibilayer liposome; ○ DPPC sonicated vesicles; △ DMPC sonicated vesicles.

diffusion is a conceptual one which expresses the average distribution of chains around the probe.

The wobbling diffusion constant $D_w$ is determined by the frequency of collisions between the probe and hydrocarbon chains, and by the mean free rotational angle between successive collisions. As the density of the hydrocarbon chain region is not so much affected by the transition, the sharp increase in $D_w$ near $T_t$ should be ascribed to the change in the frequency of collisions. The temperature dependence of $D_w$ at temperatures higher than $T_t$ is almost the same for both lipids.

*Viscosity in the cone*

From the wobbling diffusion constant $D_w$, the "viscosity in the cone," $\eta_c$, can be estimated using the same relation between the rotational diffusion constant of an ellipsoid and the viscosity of the solvent:

$$D_w = \frac{kT}{6\eta_c V_e f} \qquad (6)$$

where $V_e$ and $f$ denote the effective volume and the shape factor of the probe, $k$ the Boltzmann constant, and $T$ the absolute temperature. The viscosity in the cone represents merely the friction against the wobbling motion in the cone, and should be distinguished from the "microviscosity" which was estimated from the steady-state anisotropy $r^s$ without separating the rate and the angular range of the wobbling motion. The $V_e f$ value of DPH in hydrocarbon chains of lipids was obtained from the observed $r(t)$ of DPH in liquid paraffin because of the similar molecular properties of both solvents. The estimated values of $\eta_c$ were less than a few poise even in the ordered state, and were an order of magnitude smaller than the "microviscosity" values estimated for the same system (4). More fluid-like properties of bilayers are revealed by this analysis.

## 3. Effect of Double Bonds

In order to investigate the effect of double bonds on the dynamic properties of the hydrocarbon region, we studied (20) the fluorescence anisotropy of DPH in multibilayer liposomes of various unsaturated lecithins: DPPC, dioleoyl phosphatidyl choline (DOPC), 1-palmitoyl-2-oleoyl phosphatidyl choline (POPC), 1-palmitoyl-2-linoleoyl phosphatidyl choline (PLPC), and 1-palmitoyl-2-archidonoyl phosphatidyl choline (PAPC).

### 1) Steady-state fluorescence anisotropy

The temperature dependence of the steady-state fluorescence anisotropy $r^s$ of DPH in these multibilayer liposomes was determined within the range of 10–50°C (Fig. 4).

In this study, the effect of the first double bond was examined by a comparison between the most commonly occurring lecithin, POPC, and the disaturated lecithin, DPPC, which has a phase transition temperature ($T_t$) of 41°C. A large decrease upon the introduction of the first double bond was observed even at a temperature of 10°C, sufficiently lower than $T_t$ of DPPC. This shows that the order was substantially decreased by the first double bond. Another common unsaturated lecithin of natural membranes is PLPC. The difference between POPC and PLPC lies in the extra double bond in the hydrocarbon chain esterified at position 2. The results for POPC and PLPC were very similar over the temperature range examined, although the values of $r^s$ for PLPC were slightly smaller than those for POPC. The lecithin PAPC was stud-

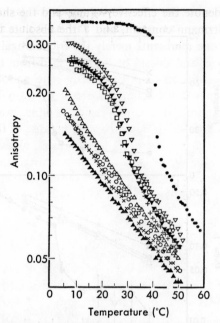

Fig. 4. Temperature dependence of the steady-state fluorescence anisotropy ($r^s$).
Single-component multibilayer liposomes: ● DPPC; △ POPC; ○ PLPC; × DOPC;
▲ PAPC. Two component liposomes (1:1 mixtures): ▽ POPC/DPPC; + DOPC/
DPPC; ▼ PAPC/DPPC; □ PLPC/DPPC.

ied since arachidonic acid contains four double bonds, and since it has
been implicated in many cell processes. The results show lower values
for $r^s$ compared to those for POPC and PLPC. The differences are, however, rather small. The lecithin DOPC was investigated to compare the
effects of two double bonds on a single hydrocarbon chain of lecithin
(PLPC) with one double bond on each chain (DOPC). The $r^s$ plots against
temperature for PLPC and DOPC revealed no difference in the temperature range of 10–50°C.

*2) Wobbling parameters*

The effect of double bonds on the dynamic structure of the hydrocarbon region was explained more clearly by the changes in the cone
angle $\theta_c$ and the wobbling diffusion constant $D_w$ shown in Fig. 5. The
large increases in the rate and angular range are produced by the first
double bond in a phospholipid at the temperatures lower than $T_t$ of the
disaturated lecithin DPPC. The values of $\theta_c$ and $D_w$ of unsaturated

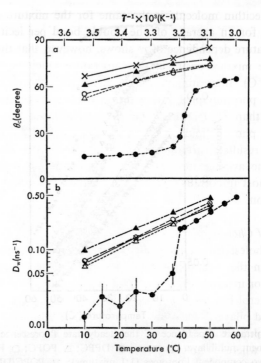

Fig. 5. a) Temperature dependence of the cone angle ($\theta_c$). b) Temperature dependence of the wobbling diffusion constant ($D_w$). Except where shown, the size of the errors in the measurements are approximately equal to the size of the symbols. The data points at 10, 25, 37, and 50°C represent the mean of at least three separate experimental values. For an explanation of the symbols, see the legend for Fig. 4.

lecithins definitely indicate the disordered state of the hydrocarbon region within the range of 10–50°C. The effect of the second and subsequent double bonds is relatively small; $\theta_c$ and $D_w$ increase in proportion to the number of double bonds, at least from the first to the fourth bond. The effect of the double bond position was revealed in the cone angle $\theta_c$: DOPC gave a value greater than PLPC and even greater than PAPC, though the values of $D_w$ for DOPC and PLPC were essentially identical.

### 3) Two component liposomes

The temperature dependence of the steady-state fluorescence anisotropy $r^s$ of two component liposomes is shown in Fig. 4. When DPPC and DOPC are mixed in a 1 : 1 ratio, the average number of double bonds per lecithin molecule is the same for the mixture as is found in POPC (*i.e.*, for an average of one double bond per lecithin molecule). The temperature dependence of $r^s$ shows, however, that the phase transition for the mixture occurs at a temperature much higher than $T_t$ of POPC ($-5°C$). Interestingly, $T_t$ for all these mixtures of DPPC with unsaturated phospholipids falls within a narrow range (25–35°C), somewhat lower than $T_t$ of DPPC, whereas $T_t$ of liposomes of corresponding unsaturated phospholipids lies within a much broader range lower than 0°C. These results clearly suggest that the fraction of unsaturated lecithin, not the average number of double bonds per lecithin molecule, plays the most important role in the dynamic properties of the hydrocarbon region.

## 4. Effect of Cholesterol

Since cholesterol is a major lipid component of many biomembranes, its effects on the dynamic structure of lipid bilayers should be an important factor in various membrane processes.

Cholesterol has been reported to have a "dual effect on fluidity" of phospholipid bilayers. The ordered structure of lipid hydrocarbon chains below the phase transition temperature is fluidized by the addition of cholesterol, whereas cholesterol reduces the fluidity above the transition temperature (*22, 23*). The lipid phase transition disappears at sufficiently high concentrations of cholesterol (*24*).

### 1) Wobbling diffusion of DPH

We studied the effect of cholesterol on the molecular motion of

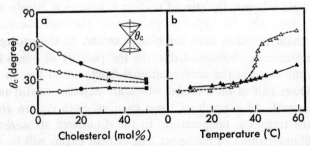

Fig. 6.  a) The effect of cholesterol on the cone angle ($\theta$) at different temperatures of DPH in DPPC-cholesterol vesicles. —— 10.5; --- 40; —— 49.5°C. Cholesterol concentrations are: △ 0; ○ 10; ● 20; ▲ 33; and ■ 50 mol%. b) Temperature dependence of the cone angle of DPH for 33 mol% cholesterol (▲) and for pure DPPC vesicles (△).

lipid hydrocarbon chains in DPPC-cholesterol vesicles (9). Again DPH
was used to probe its dynamic environment. The anisotropy decay of DPH
was biphasic, and its motion was adequately described by the wobbling
in the cone model. The effect of cholesterol on the cone angle $\theta_c$ is illus-
trated in Fig. 6. In the disordered state at 49.5°C, the incorporation of
cholesterol decreased $\theta_c$ sharply, whereas it increased $\theta_c$ slightly in the
ordered state at 10.5°C. An intermediate result was obtained in the phase
transition region at 40°C. The cone angle appears to converge to 30°,
irrespective of temperature, at sufficiently high concentrations of cholesterol (Fig. 6a). In the presence of 33 mol% cholesterol, $\theta_c$ increased
gradually as temperature was raised (Fig. 6b). The abrupt increase in
$\theta_c$ at the phase transition of pure DPPC vesicles was completely suppressed by cholesterol.

The relaxation time, $\phi$ in Eq. (4), was reduced to less than 0.6 ns by
cholesterol (33 mol%) over the temperature range 10–60°C. On account
of the rather scattered values of $D_w$, only a qualitative conclusion was
deduced: $D_w$ increases appreciably below the transition temperature,
while $D_w$ changes little above it.

*2) Several effects of cholesterol*

Thus, the effect of cholesterol on the cone angle would be called the
"dual effect," while the wobbling rate did not show any "dual effect."
Therefore, the reported changes in the fluidity should reflect mainly the
change in the cone angle $\theta_c$. The differential effect of cholesterol on $\theta_c$
and $D_w$ may be explained by a difference between two parts of cholesterol
molecules: a bulky rigid steroid nucleus and a relatively narrow flexible
tail. In lipid bilayers, the steroid nucleus is immersed largely in the lipid
chain region, with its hydroxyl group near the water surface and its
methyl chain terminus near the bilayer center. In the ordered state at
low temperatures, cholesterol disturbs the packing of lipid chains, and
may chiefly increase the conformational freedom and the wobbling rate
of the lower half of lipid chains which are adjacent to the narrow tails
of cholesterol. As a result of these changes, both $\theta_c$ and $D_w$ increase
below the transition temperature. In the disordered state, the range of
the wobbling motion of the upper half of lipid chains will be suppressed
by the steroid part. Therefore the cone angle decreases. Little increase
in $D_w$ may reflect the fact that the wobbling rate of DPH depends
strongly on the wobbling rate of the lower half of lipid chains.

*3) Fluorescence lifetime*

A single exponential decay of fluorescence intensity of DPH was observed over the cholesterol content 0–50 mol%. This finding indicates that there is no distinct phase separation in DPPC-cholesterol vesicles, for all DPH have a single lifetime showing similar environments.

## II. PROTEIN-LIPID INTERACTION: THE CASE OF CYTOCHROME OXIDASE

One of the important problems regarding membranes is the interaction between membrane proteins and lipids. The effect of lipids on the conformation and function of membrane proteins and the influence of membrane proteins on the physical state of neighboring lipids were studied in reconstituted vesicles of purified cytochrome oxidase and synthetic phospholipids using the fluorescence technique.

Cytochrome oxidase is the terminal enzyme of the respiratory chain catalyzing the transfer of electrons from cytochrome $c$ to molecular oxygen. It makes a multipeptide transmembrane complex containing heme $a$ groups in mitochondrial inner membrane. The size of the enzyme complex has been estimated to be about 5 nm × 6 nm in the plane of the membrane, and 11 nm for the vertical direction (*25, 26*).

### 1.  Conformational Changes in Cytochrome Oxidase

Temperature-induced conformational changes in oxidized cytochrome oxidase were investigated in phosphate buffer containing Emasol (ECP) and in three lecithin vesicles with different transition temperatures, DPPC (41°C), DMPC (23°C), and DOPC (−20°C) (*27, 28*).

To detect the conformational changes in the protein, fluorescent probe N-(1-anilinonaphthyl-4) maleimide (ANM) was labeled to the enzyme. ANM reacts specifically with SH groups of proteins. Most of the added ANM reacted with subunit I of cytochrome oxidase though it has seven SH groups per heme $a$. The fluorescent probe DPH was used to detect the physical state of hydrocarbon chains.

*1) Temperature dependence of fluorescence intensity and emission anisotropy*

The steady-state fluorescence intensity, $I_T$, and the emission anisotropy, $r^s$, of ANM bound of cytochrome oxidase and of DPH incorpo-

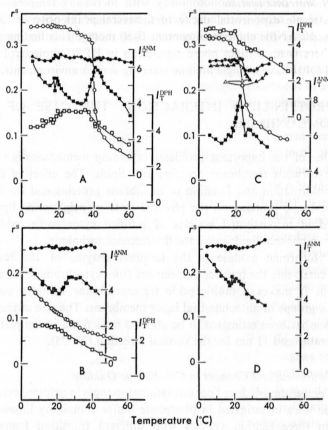

Fig. 7.   Temperature dependence of the fluorescence intensity, $I_T$ (■ ANM; □ DPH), and the steady-state fluorescence anisotropy, $r^s$ (● ANM; ○ DPH). A) DPPC-enzyme. B) DOPC-enzyme. C) DMPC-enzyme (● $r_1$ curve of ANM; ▲, △ $r_2$ curve of ANM). D) ECP-enzyme.

rated in the hydrocarbon region of vesicles are plotted against temperature in Fig. 7. The phase transition of lipid hydrocarbon chains was detected by the sharp changes in $r^s$ of DPH for DPPC and DMPC vesicles containing cytochrome oxidase. The transition temperature of these vesicles was the same as that of pure lecithin vesicles.

The fluorescence intensity of a probe is sensitive to the solvent, and is generally quenched by collision with surrounding molecules. Since collision is more frequent at higher temperatures, the fluorescence intensity will decrease monotonously with increasing temperature when there are no other quenching factors. Several peaks observed in the $I_T$ curves of ANM indicate, however, that there are other mechanisms which increase or decrease the intensity at a particular temperature. One probable mechanism of such temperature-dependent peaks is the change in the efficiency of excitation energy transfer from ANM to heme $a$ of cytochrome oxidase. The emission spectra of ANM bound to the enzyme overlap partially with the strong Soret absorption band of oxidized heme $a$. ANM is fairly immobilized in cytochrome oxidase judging from the high values of $r^s$ (about 0.25) as compared with its limiting anisotropy of 0.365. The orientation between the emission moment of ANM and the absorption moment of heme $a$ should thus be restricted. Therefore, if a conformational change in cytochrome oxidase involving a rearrangement between ANM and heme $a$ occurs, one might expect a change in the fluorescence intensity. Since ANM fluorescence intensity is sensitive to the solvent polarity, it is also possible that the fluorescence intensity is altered by the conformational change in cytochrome oxidase which leads to a change in the polarity around ANM.

*2)   Conformational change induced by lipid phase transition*

A large peak in the $I_T$ curve of ANM observed at the phase transition temperature of DPPC or DMPC vesicles containing cytochrome oxidase may be assigned to the conformational change induced by the lipid phase transition. The change in the $I_T$ curve can, in principle, be interpreted by a change in the polarity around ANM without any intramolecular structural change. The sharp increase and subsequent sharp decrease in the $I_T$ curve near the transition temperature may be, however, difficult to explain as solely due to an environmental change induced by the transition.

*3)   Intrinsic conformational change in cytochrome oxidase*

All present samples containing cytochrome oxidase showed peaks or sharp changes in the $I_T$ curves of ANM at 17–20°C. The $r^s$ curves of DPH in this temperature range were very smooth, and indicate that DPPC and DMPC are in the ordered state and DOPC is in the disordered state. Therefore, these peaks in the $I_T$ curves of ANM could not be attributed to the lipid phase transition.

The peaks would be due to the intrinsic conformational change in cytochrome oxidase. The inflections in the $I_T$ curves of DPH in this temperature region suggest also the presence of a conformational change in the enzyme which may influence the rate of energy transfer between DPH and heme $a$.

### 4) Temperature dependence of activity

The activity of the cytochrome oxidase was determined in these lecithin vesicles and in ECP, by measuring the rate of oxygen uptake in the presence of ascorbate and cytochrome $c$ (29). A polaro-graphic oxygen sensor was used to monitor the oxygen concentration. A break in the Arrhenius plot was observed at about 20°C for the enzyme in ECP and was observed at about 23°C for the cytochrome oxidase-DOPC system. These results indicate that the intrinsic conformational change appears near these temperatures.

### 2. Effect of Cytochrome Oxidase on Lipid Dynamics

The effect of the protein-lipid interaction on the dynamic properties of the hydrocarbon region was investigated in a simple system composed of purified cytochrome oxidase and synthetic phospholipids, DMPC or DOPC (30).

### 1) Steady-state fluorescence anisotropy

The steady-state fluorescence anisotropy, $r^s$, of DPH was measured in cytochrome oxidase-DMPC vesicles. In pure DMPC vesicles, $r^s$ decreased sharply at about 23°C, the phase transition temperature of DMPC bilayers. Below the phase transition, the addition of cytochrome oxidase had little effect on $r^s$. In the disordered phase above the transition temperature, on the other hand, the presence of cytochrome oxidase greatly increased $r^s$. The effect was basically proportional to the oxidase/lipid ratio, and at the molar ratio at 4/100 the phase transition was no longer observed. A very similar temperatures profile has been reported for bacteriorhodopsin-DMPC vesicles (31). The effect of cytochrome oxidase on DOPC vesicles is similar to that on DMPC in the disordered state.

### 2) Wobbling motion

The difference between the effect of cytochrome oxidase on the ordered and that on the disordered state of lipid chains was examined using nanosecond fluorescence depolarization measurements at 10 and 35°C, below and above the transition temperature of DMPC. The observed biphasic decay of fluorescence anisotropy, $r(t)$, of DPH was analyzed in terms of wobbling in the cone model. The results are shown in Fig. 8. The effect of oxidase on the wobbling motion of DPH was

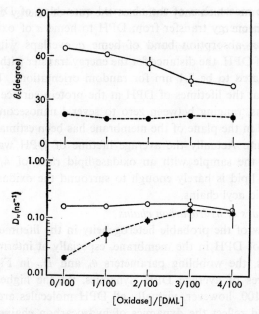

Fig. 8. The wobbling diffusion constant, $D_w$, and the cone angle, $\theta_c$, for the wobbling motion of DPH in oxidase-DMPC vesicles. ○ at 35°C; ● at 10°C. Each point is an average over two to four samples. The vertical bars represent standard deviation; where not indicated, the deviation was smaller than the size of the symbol.

markedly different between the two temperatures investigated. Above the phase transition, the addition of the protein progressively reduced $\theta_c$ while $D_w$ remained the same. Below the phase transition, on the other hand, the addition of the protein increased $D_w$ while $\theta_c$ remained the same. On the whole, the effect of cytochrome oxidase was to converge $\theta_c$ to the value of the ordered state, and $D_w$ to the value of the disordered state, irrespective of temperature. This result is qualitatively similar to the effect of cholesterol, as described before.

*3)   Fluorescence lifetime*

Furthermore, the fluorescence lifetime of DPH in vesicles was affected by the addition of cytochrome oxidase. In pure DMPC vesicles, the total fluorescence intensity, $I(t)$, decayed almost single exponentially both below and above the phase transition. Cytochrome oxidase even at the molar ratio of 1/100 caused marked deviation from a single exponential decay and greatly reduced the average lifetime. The most conceivable mechanism of the observed quenching of DPH fluorescence is excitation energy transfer from DPH to heme $a$ of oxidase. Since the strong Soret absorption band of heme $a$ overlaps with the emission spectrum of DPH, the distance for the energy transfer with 50% efficiency, $R$, is estimated to be 5.1 nm for random orientations. This value of $R$ suggests that the lifetimes of DPH at the protein surface are distributed around 1 ns, ranging between zero to several nanoseconds. The size of the oxidase in the plane of the membrane has been estimated to be about 5 nm × 6 nm. Actually the average lifetime of DPH was about 1.4 ns (35°C) for the sample with an oxidase/lipid ratio of 4/100, where the amount of lipid is barely enough to surround the oxidase with a single layer of lipid acyl chains.

### 4) Dynamics of hydrocarbon chains

In view of the probable heterogeneity in the lifetimes and motional properties of DPH in the membrane, especially at intermediate oxidase/lipid ratios, the wobbling parameters $\theta_c$ and $D_w$ in Fig. 8 should be gross indices of average DPH dynamics. At the highest oxidase/lipid ratio of 4/100, however, virtually all DPH molecules are at the protein surface, and reflect the dynamics of hydrocarbon chains in the vicinity of the protein.

Here it should be noted that the angular range $\theta_c$ refers to the motion in the nanosecond time range and does not necessarily deny the presence of large-angle rotations in the micro- and millisecond time range. Also, the axis of the wobbling motion does not necessarily coincide with the normal of the membrane.

Electron spin resonance (ESR) studies using spin-labeled fatty acid or phospholipid have shown that cytochrome oxidase is surrounded by a layer of immobilized lipid hydrocarbon chains. The exchange between these "boundary lipids" and bilayer lipids takes at least 50 ns (32, 33). Deuterium nuclear magnetic resonance ($^2$H-NMR) studies (34–36) for membranes containing cytochrome oxidase detected only one spectral component of deuterated phospholipid above the transition temperature. This fact implies that exchange rate between the boundary and bilayer lipids exceeds $10^3$ s$^{-1}$. Moreover, the protein disorders the lipid hydrocarbon chains both below and above the phase transition.

The results of our fluorescence measurements as well as the results of ESR and NMR studies are accounted for in a unified manner by two types of motion differing in their time range. In the nanosecond range, hydrocarbon chains of lipids in the bilayer region exhibit a fast wobbling motion. The wobbling motion of DPH reflects this chain motion. The order parameter for the bilayer lipid obtained by ESR relates to $\theta_c$ of the wobbling motion. In the vicinity of the protein, $\theta_c$ of hydrocarbon chains is severely reduced by the restriction of the irregular surface of protein, though the axis of the wobbling motion differs from chain to chain. The immobilization observed by ESR reflects this reduction in $\theta_c$. In the microsecond range, boundary lipid molecules exchange their places on the protein surface by thermal energy that overcomes the depth of the potential well. The exchange motions of lipids on the protein surface are not observed by fluorescence depolarization nor by conventional ESR spectroscopy, both of which are insensitive to microsecond re-orientations. The NMR order parameter, on the other hand, relates to the averaged direction of a probe over the time range of microseconds. Thus, protein reduces the NMR order parameter, for it reflects the chain direction in all types of cones on the irregular protein surface.

The rate of chain wobbling, or $D_w$, in the bilayer below the phase transition is extensively reduced by cooperative interaction between chains (Fig. 3). Introduction of a rigid molecule such as protein and cholesterol into the bilayer should disturb the cooperative action and increase the rate of chain motion in its neighborhood. The observed increase in $D_w$ caused by the addition of oxidase below $T_t$ may be attributed to this effect.

## III. CHROMOPHORE IN BACTERIORHODOPSIN

The purple membrane of *Halobacterium halobium* consists of a specific protein, bacteriorhodopsin, and several lipids. Bacteriorhodopsin binds retinal chromophore stoichiometrically *via* a protonated Schiff base linkage to a lysine residue.

Light absorption by the chromophore initiates a photochemical cycle, by which protons are actively transported from the cytoplasm to the outer side, and thereby an electrochemical proton gradient is produced across the cell membrane (37).

In the purple membrane, bacteriorhodopsin forms an almost perfect two-dimensional crystal with $P_3$ symmetry. A three-dimensional model of the protein structure has been proposed from the electron diffraction analysis for tilted membrane specimens (38). In this model the protein consists of seven $\alpha$-helices of about 40 Å long each of which penetrates the membrane almost perpendicularly.

Information about the location and movement of the chromophore in the protein is obscure yet, though it is important for understanding the molecular mechanism of the light-induced proton pump.

## 1. Immobility of the Chromophore in the Binding Site

Earlier studies of transient absorption dichroism of the purple membrane have shown that the rotational motions of the protein in the membrane are absent in the time range of $10^{-4}$–$10^{-3}$s (39–41). However, the reported values of absorption anisotropy are considerably lower than the theoretical maximum. This suggests the possibility that the chromophore and/or the protein have a freedom of rotation within the membrane faster than the microsecond time range. To clarify this point, we studied (42) the rotational motions in the two different time ranges, the nanosecond time range was studied by fluorescence depolarization measurements and the millisecond time range by transient absorption dichroism measurements.

In the former fluorescence depolarization measurements, reduced purple membrane was used in which the chromophores were converted to fluorescent derivatives but the hexagonal crystalline structure was maintained (37). Purple membrane fragments were reduced with NaBH$_4$ under illumination with visible light. The reduced products, bR$^{red}$, were then irradiated with ultraviolet (UV)-light until the characteristic absorption spectrum of UV-converted form of bR$^{red}$ fully developed.

### Immobilization

A remarkably high value of time-averaged fluorescence anisotropy of the UV-converted form of bR$^{red}$ close to the theoretical maximum of 0.4, 0.385±0.005 (SD for 8 samples), was observed at 10°C, and did not change detectably when the temperature was varied from 4 to 50°C.

As the chromophore in the UV-converted form is all-*trans* retro-retinyl presumably attached to the same lysine residue of the original bacteriorhodopsin, its transition moments are expected to lie along the original polyene chains. The fact that the anisotropy remained at a high constant value after the pulsed excitation (Fig. 9) indicates the absence of the rotational motion of the chromophore, at least in the ns time range. The slight difference between the observed anisotropy of 0.385 and the theoretical maximum of 0.4 could be ascribed to a small

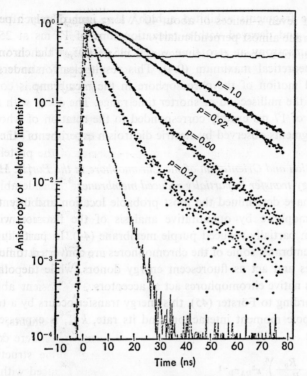

Fig. 9. Experimental decay curves of the total fluorescence intensity $S^{ex}(t;p)$ of the UV-converted form at various values of $p$ (filled circles). The chain line shows the response function $g(t)$. The intensity decay curve at $p=1$ could be described with a single-exponential function with a decay constant of 21.0 ns. The fluorescence anisotropy decay curve of the UV-converted membrane (at $p=1$) is also shown (zigzag line). Excitation was at 381 nm and emission above 460 nm was collected together. Solvent: 67% (v/v) glycerol (pH 7.0) at 10°C.

angular deviation (about 9°) between the absorption and the emission transition moments. Another probable explanation of the difference, however, is the effect of scattering of the excitation and fluorescence lights in the sample.

The complete immobilization of the chromophore in purple membrane was confirmed by measurements of transient absorption dichroism. Due to the photochemical reaction induced by laser flash irradiation, the absorption band of bacteriorhodopsin changes via several intermediates of the photochemical cycle. The transient absorption of purple membrane fragments exhibited a high value immediately after the ir-radiation, and decayed with a relaxation time of 17 ms at 25°C. The limiting anisotropy at zero time was $0.395\pm0.005$, which is very close to the theoretical maximum (0.4). This result clearly shows that the rotational motion of the chromophore in the membrane is completely absent in the millisecond or shorter time range. The decay with a relaxation time of 17 ms exactly corresponded to the rotation of whole membrane fragments observed by electric dichroism experiments (43).

## 2. Location and Orientation of the Chromophore in the Purple Membrane
### 1) Energy transfer in partially reduced membranes

We have determined the most probable location and orientation of the chromophore by quantitative analysis of the fluorescence energy transfer in partially reduced purple membrane (44). In partially reduced purple membrane, some of the chromophores are converted to fluorescent derivatives and act as fluorescent energy donors, while the others that remain as native chromophores act as acceptors.

According to Förster (45), this energy transfer occurs by a transition dipole-dipole moment interaction, and its rate, $k_{DA}$, is expressed by the equation:

$$k_{DA}=\left(\frac{R_0}{R_{DA}}\right)^6 \kappa^2{}_{DA}\tau_D^{-1} \tag{7}$$

where $R_{DA}$ is the distance between the center of the transition moments of the donor D and acceptor A, and $\tau_D$ is the fluorescence lifetime of the donor in the absence of the acceptor. $\kappa_{DA}$ is the orientation factor defined by the equation:

$$\kappa_{DA}=e_A\cdot e_D - 3(e_A\cdot R_{DA})(e_D\cdot R_{DA}) / (R_{AD})^2 \tag{8}$$

where $e_A$ and $e_D$ are unit vectors of the emission dipole moment of D and the absorption dipole moment of A, respectively. The value of $R_0$ (in cm) is given from the spectroscopic data.

Because of the symmetry of the two-dimensional arrangement of bacteriorhodopsin in purple membrane, the location and orientation of all chromophores are uniquely determined when we assume the position and orientation of the chromophore in the unit cell. Thus, the rate of energy transfer from a donor to any acceptor in the same membrane can be expressed as a function of three coordinates $(y, \xi, \phi_A)$ and an angle

$\Delta\phi$ (in the membrane plane) between the emission dipole moment of the donor and the absorption dipole moment of its original chromophore. Two coordinates, $y$ and $\xi$, express the position of the chromophore in the unit cell and coordinate $\phi_A$ expresses the direction of the absorption dipole moment of the native chromophore in the unit cell. Furthermore, if the reduction occurs in a random way, the probability of finding an acceptor at any chromophore site should be proportional to $1-p$ where $p$ is the degree of reduction. Then, the fluorescence decay (ensemble average) of the donor is given by:

$$s(t;p)=\exp(-t/\tau_D)\Pi_j[p+(1-p)\exp(-k_jt)] \tag{9}$$

where $k_j$ is the rate of energy transfer to an acceptor at site $j$. At times much shorter than $\tau_D$, the fluorescence decay can be approximated as follows:

$$s(t;p)\simeq 1-\Gamma(p)t. \tag{10}$$

The average rate of deactivation of the excited state $\Gamma(p)$ is given by:

$$\Gamma(p)=(1-p)\Sigma_j k_j+\frac{1}{\tau_D}. \tag{11}$$

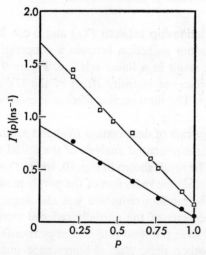

Fig. 10. The average rate, $\Gamma(p)$, of deactivation of the excited state of the donor as a function of $p$, for the non-UV-converted form ($\square$) and for the UV-converted form ($\bullet$). The initial time region (from $-2$ to $+10$ ns in Fig. 9) of the fluorescence decay curve was analyzed with a double-exponential function.

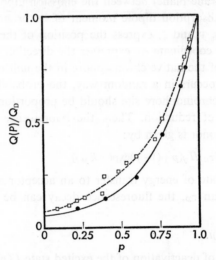

Fig. 11. The relative fluorescence quantum yield, $Q(p)/Q_D$, of the donor as a function of $p$, for the non-UV-converted form ($\square$) and for the UV-converted form ($\bullet$). Here, $Q_D$ is the absolute quantum yield observed at $p=1$. Excitation was at 370 nm ($\square$) or at 381 nm ($\bullet$). Solid line shows the calculated curve $Q^{cal}(p)$ for the most probable location and orientation of the chromophore.

Thus the linear relationship between $\Gamma(p)$ and $p$ can be used to test the assumption of random reduction because a cooperative or nonrandom reduction will not result in a linear relation. Figure 9 shows the decay curves of the fluorescence intensity $s(t;p)$ of the UV-converted form at various values of $p$. The fluorescence lifetime increases with the increase in the degree of reduction $p$. In order to examine the linear relationship between the average rate of deactivation $\Gamma(p)$ and $p$, the value of $\Gamma(p)$ was obtained by double-exponential analysis of the initial fluorescence decay curves ($\leq 10$ ns). The result shown in Fig. 10, where $\Gamma(p)$ is plotted against $p$, indicates clearly that the reduction of the purple membrane occurred in a random way. The random reduction was also suggested from circular dichroism measurements of the partially reduced membrane. The slope of $\Gamma(p)$ in Fig. 10 gave the total rate of energy transfer from a donor to acceptors at all other sites, $\Sigma_j k_j$. Fluorescence quantum yields were measured under the same solvent conditions as used for fluorescence decay measurements. The relative quantum yields with respect to the value obtained at $p=1$ are plotted against $p$ in Fig. 11.

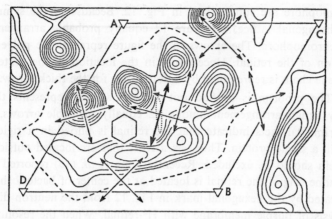

Fig. 12.   The most probable location and orientation of the chromophore in bacteriorhodopsin. Among the 12 possible arrangements (arrows) of the polyene chain of
the retinal, the most likely is represented by the thick arrow. The length (12 Å) of the
arrow corresponds to that of the polyene chain within the membrane plane. Uncertainty in the location of the center is shown by the dotted line and uncertainty in the
orientation is about 20°. The hexagonal mark in the figure shows the position of the
$\beta$-ionone ring portion of the retinal that was determined by King *et al.* (*46*). The electron scattering density map was taken from the results reported by Unwin and Henderson (*38*).

## 2)   Location in the membrane plane

The absolute quantum yield of the UV-converted form at $p=1$ was
0.24. By integrating Eq. (9), the fluorescence quantum yield of the donor
$Q(p)$ at any value of $p$ can be calculated as a function of three coordinates
$y$, $\xi$, $\phi_A$, and the angle $\Delta\phi$. Therefore, finding a set of variables ($y$, $\xi$,
$\phi_A$, $\Delta\phi$) that can satisfactorily describe the experimental data of $\Sigma_j k_j$
and $Q(p_n)$ at $p=p_n$ ($n$ is data number) enables us to determine the most
probable location and orientation of the chromophore in bacteriorhodopsin. Because of the practical difficulty in determining the values of four
independent variables at the same time, three parameters $y$, $\xi$, $\phi_A$, were
determined by the least squares method using a given value of $\Delta\phi$. The
procedure was repeated for different values of $\Delta\phi$. Discrepancy between
the experimental and calculated values $Q(p_n)$ appeared when values
larger than 20° were given for $\Delta\phi$. The locations ($y$, $\xi$, $\phi_A$) determined
for different values of $\Delta\phi$ less than 20° were, however, much the same;
the uncertainty in $\Delta\phi$ affected the estimation very weakly. The final
results obtained on the most probable location and orientation of the

chromophore are summarized in Fig. 12. Because of the symmetry of the hexagonal lattice, there are 12 equally probable arrangements of the chromophore. The arrows in the figure represent the polyene chain portion of the retinal. Uncertainty in the location of the center of the polyene chain is represented by a dotted line for the thick arrow.

The problem is to find out the arrow that represents the most likely actual arrangement among 12 equally probable arrows. Several experimental facts indicate that the retinal is buried in a rigid pocket within a single protein. Thus, six arrows that are located outside of the protein should be excluded. Recently King et al. (46) reported that the $\beta$-ionone ring of the retinal is located at the center of bacteriorhodopsin, indicated by the hexagonal mark in Fig. 12 from the neutron diffraction analysis of purple membrane with $H^2$-retinal. When the resolution (8.5 Å) of their Fourier synthesis map and the experimental error in our data are taken into consideration, the present result is not incompatible with their result. Three or four arrows inside the protein could be fitted to their result with only a slight movement or reorientation of arrows. Further information is necessary to clarify the location and orientation.

### 3) Transmembrane location of the retinal chromophore

We have also applied the fluorescence energy transfer technique to the determination of transmembrane location of the retinal chromophore in the purple membrane. A theoretical consideration predicts that the efficiency of the energy transfer from a donor to rapidly diffusing acceptors can be given as a function of the most accessible distance between the donor and acceptors. In this study, the fluorescence intensity decay of the completely reduced and then UV-converted purple membrane was measured in the presence of cobalt-ethylenediamine tetraacetate (Co-EDTA) dissolved in solution (pH 8.9). As Co-EDTA has a weak absorption band around 460 nm ($\varepsilon_{max} = 18$ cm$^{-1}$M$^{-1}$), it is available as an acceptor of excitation energy of the fluorescent retinal derivative.

When the concentration of Co-EDTA was increased up to 0.25 M, the average fluorescence lifetime of the donor was decreased from 20.5 to 18.6 ns (20°C). No detectable change in the fluorescence lifetime was induced upon addition of 0.25 M calcium(Ca)-EDTA which has no absorption band in the visible region. From this result, the most accessible distance, $D$, was estimated to be about 12 Å. Since the transfer efficiency was unaffected by the addition of salt (1 M NaCl), no special attention was given to the electrostatic interaction between Co-EDTA and the membrane surface. Then, the distance $D$ can be approximated by a sum of the radius of Co-EDTA and the depth of the chromophore from the membrane surface.

Transmembrane location of the chromophore has been also determined from the neutron diffraction study (47). The result shows that the retinal is situated about 17 Å below the membrane surface, somewhat different from our result.

## IV.  DYNAMIC STRUCTURE OF BIOLOGICAL MEMBRANES

We studied the dynamic structure of the membrane interior, as probed by DPH, of various biological membranes with known compositions (48). The results were compared with those obtained in the model membranes described above.

*1)  Steady-state fluorescence anisotropy*

The temperature dependence of the steady-state fluorescence anisotropy $r^s$ of DPH was measured in four biological membranes: human erythrocyte membrane (E), rabbit sarcoplasmic reticulum membrane (S), rat liver mitochondrial membrane (M), and the purple membrane of *H. halobium* (P). No apparent break in the temperature dependence of $r^s$ indicating transition was observed.

*2)  Wobbling motion of DPH in membranes*

Nanosecond fluorescence depolarization measurements were made for the four membranes at two temperatures, 10 and 35°C. As seen in the typical case in Fig. 1b, $r(t)$ in all cases was biphasic, and was interpreted in terms of wobbling in the cone model. Because of the probable heterogeneity in a biological membrane, the wobbling parameters, $\theta_c$ and $D_w$, thus estimated should be considered as averages over various motions of DPH in the membrane. The present results on the biological membranes together with the wobbling parameters of DPH in lipid bilayers and reconstituted membranes described before are summarized ig Fig. 13.

*3)  Temperature dependences of $D_w$ and $\theta_c$*

The first thing to note in Fig. 13 is that in spite of the large differences in compositions of the four biological membranes as will be discussed below, temperature dependences of $D_w$ and $\theta_c$ in the membranes are quite similar. Thus, when the temperature was raised from 10 to 35°C, $D_w$ increased about 3-fold, corresponding to activation energies of 7–8 kcal/

A. IKEGAMI ET AL.

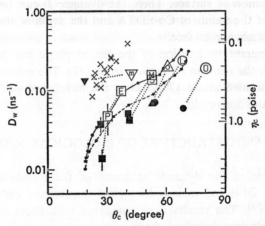

Fig. 13. Wobbling parameters of DPH in model and biological membranes. Solid
curve, sonicated vesicles of DPPC from 5°C (lower left) to 55°C (upper right); dashed
curve, sonicated vesicles of DMPC from 5 to 50°C. Data points on the curves (small
closed circles) are 5°C apart, except in the transition region where the intervals are
2–3°C. Crosses, sonicated DPPC vesicles containing 33 mol% cholesterol between
10 and 60°C. Where two large symbols are connected with a dotted line, the open
symbol with a letter in it denotes $D_w$ and $\theta_c$ at 35°C or 37°C, and the closed symbol at
10°C. P in square, purple membrane; E in square, erythrocyte membrane; M in square,
mitochondrial membrane; S in square, sarcoplasmic reticulum membrane; L in circle,
multibilayer liposomes of PLPC; O in circle, cholate dialysis vesicles of DOPC, m
in inverted triangle, cholate dialysis vesicles of DMPC containing cytochrome oxidase
at a heme $a$/lipid molar ratio of 3/100; o in triangle, cholate dialysis vesicles of DOPC
containing cytochrome oxidase at a heme $a$/lipid molar ratio of 2/100. The right-hand
scale for $\eta_c$ is an approximate one.

mol. At the same time, $\theta_c$ increased by about 10° in all membranes except
the purple membrane. The same temperature dependences of $\theta_c$ and $D_w$
were found in many model systems that contain unsaturated lecithins.

The similarity of the temperature dependences of $\theta_c$ and $D_w$ is easily
seen, in the parallelism between dotted lines in Fig. 13, where dotted
lines connect $D_w$ and $\theta_c$ at 10°C (closed symbols) and those at 35°C (open
symbols with letters). Saturated lecithins above the phase transition also
gave activation energies for $D_w$ of 8–9 kcal/mol. These findings suggest
an important role of unsaturated phospholipids in the motional properties of biological membranes.

*4) Wobbling parameters*

The results in Fig. 13 suggest a strong correlation between $\theta_c$ and the composition of the membrane. The values of $\theta_c$ in pure lecithin bilayers at 35°C spread from about 20° in the ordered state of saturated lecithins to about 75° in the disordered state of unsaturated lecithins. Since cholesterol reduced $\theta_c$ dramatically in the disordered state of bilayers, the relatively small $\theta_c$ values of erythrocyte membrane can be attributed to its high cholesterol content, with a cholesterol/phospholipid ratio of 0.8 to 1.

In the purple membrane, bacteriorhodopsin molecules (75% of the membrane mass) form a crystalline lattice, with bilayer lipid in between. The rigid crystalline structure of the protein apparently reduces $\theta_c$ and $D_w$ of DPH. Addition of cytochrome oxidase in pure lecithin vesicles in the disordered state reduced $\theta_c$ whereas $D_w$ remained high. The effect of the protein is similar to that of cholesterol.

The relation between the rate of wobbling, $D_w$, and the composition of the membrane is somewhat obscure. A rough correlation between $D_w$ and $\theta_c$ is observed, however, in biological and model membranes containing unsaturated phospholipids see (Fig. 13).

As a conclusion, the dynamic structure of biological membrane represented by the motional properties in the lipid hydrocarbon chain region may be expressed, to a first-order approximation, by the following simple formula: biological membrane $\cong$ unsaturated phospholipid + protein + cholesterol. The important factor is the relative amounts of the three components, and not the compositions within each term.

## SUMMARY

Nanosecond fluorescence spectroscopy was applied to investigate the structure and dynamics of phospholipid bilayers, reconstituted membranes and biological membranes.

The molecular motion of the hydrocarbon chains in membranes was studied with the fluorescence depolarization technique using DPH as a probe. The anisotropy decay of DPH in all membranes studied was biphasic, consisting of an initial decrease and a following stational phase which indicate the restricted rotational motion of the probe. The biphasic anisotropy decay was analyzed using the wobbling in the cone model, in which the long axis of the fluorescent probe was assumed to wobble freely within a cone of semiangle $\theta_c$ with a wobbling diffusion constant $D_w$.

In the saturated phospholipid bilayers, the estimated values of $\theta_c$ are about 20° and 70° in the ordered and disordered states of lipid hydrocarbon chains, respectively. Both $\theta_c$ and $D_w$ increase sharply at the order-disorder phase transition. The values of $\theta_c$ and $D_w$ in the unsaturated phospholipid bilayers definitely indicate the disordered state within the range of 10°–50°C. The results of two component liposomes of saturated and unsaturated lipids suggest that the fraction of unsaturated lecithin, not the average number of double bonds per lecithin molecule, plays the important role in the dynamic properties of the hydrocarbon region. The incorporation of cholesterol into saturated lipid vesicles decreases $\theta_c$ in the disordered state, whereas it increases $\theta_c$ slightly in the ordered state of hydrocarbon chains.

The interaction between membrane proteins and lipids was studied in reconstituted vesicles of purified cytochrome oxidase and phospholipids. Temperature induced conformational changes in the protein were detected by the fluorescence intensity of ANM labeled to the enzyme. In addition to the intrinsic conformational change in cytochrome oxidase at about 20°C, the conformational changes induced by the lipid phase transition were observed. The effect of cytochrome oxidase on the dynamic properties of the hydrocarbon region is qualitatively similar to the effect of cholesterol. The addition of the protein progressively reduced $\theta_c$ in the disordered state while it increased $D_w$ in the ordered state.

By the quantitative analysis of the fluorescence energy transfer in partially reduced purple membrane, we determined the most probable location and orientation of the chromophore, retinal, in the purple membrane.

Finally dynamic structure of various biological membranes is discussed in terms of the wobbling parameters $\theta_c$ and $D_w$ of DPH.

## REFERENCES

1. Yguerabide, J. (1972) *Methods Enzymol.* **26**, 498–578
2. Kinosita, K., Jr., Kawato, S., & Ikegami, A. (1977) *Biophys. J.* **20**, 289–305
3. Chen, L.A., Dale, R.E., Roth, S., & Brand, L. (1977) *J. Biol. Chem.* **252**, 2163–2169
4. Kawato, S., Kinosita, K., Jr., & Ikegami, A. (1977) *Biochemistry* **16**, 2319–2324
5. Dale, R.E., Chen, L.A., & Brand, L. (1977) *J. Biol. Chem.* **252**, 7500–7510
6. Lakowicz, J.R., Prendergast, F.G., and Hogen, D. (1979) *Biochemistry* **18**, 508–519
7. Wolber, P.K. & Hudson, B.S. (1981) *Biochemistry* **20**, 2800–2810
8. Veatch, W.R. & Stryer, L. (1977) *J. Mol. Biol.* **117**, 1109–1113

9. Kawato, S., Kinosita, K., Jr., & Ikegami, A. (1978) *Biochemistry* **17**, 5026–5031
10. Sené, C., Genest, D., Obrénovitch, A., Wahl, P., and Monsigny, M. (1978) *FEBS Lett.* **88**, 181–186
11. Hildenbrand, K. & Nicolau, C. (1979) *Biochim. Biophys. Acta* **553**, 365–377
12. Sklar, L.A., Hudson, B.S., & Simoni, R.D. (1977) *Biochemistry* **16**, 819–828
13. Vanderkooi, J.M., Ierokomas, A., Nakamura, H., & Martonosi, A. (1977) *Biochemistry* **16**, 1262–1267
14. Dockter, M.E., Steinemann, A., & Schatz, G. (1978) *J. Biol. Chem.* **253**, 311–317
15. Veatch, W. & Stryer, L. (1977) *J. Mol. Biol.* **113**, 89–102
16. Fung, B. K.-K. & Stryer, L. (1978) *Biochemistry* **17**, 5241–5248
17. Fernandez, S.M. & Berlin, R.D. (1976) *Nature* **264**, 411–415
18. Thomas, D.D., Carlsen, W.F., & Stryer, L. (1978) *Proc. Natl. Acad. Sci. U.S.* **75**, 5746–5750
19. Fleming, P.J., Koppel, D.E., Lau, A.L.Y., & Strittmatter, P. (1979) *Biochemistry* **18**, 5458–5464
20. Stubbs, C.D., Kouyama, T., Kinosita, K., Jr., & Ikegami, A. (1981) *Biochemistry* **20**, 4257–4262
21. Kinosita, K., Jr., Kawato, S., & Ikegami, A. (1982) *Biophys. J.* **37**, 461–464
22. Lippert, J.L. & Peticolas, W.L. (1971) *Proc. Natl. Acad. Sci. U.S.* **68**, 1572–1576
23. Oldfield, E. & Chapman, D. (1971) *Biochem. Biophys. Res. Commun.* **43**, 610–616
24. Hinz, H.-J. & Sturtevant, J.M. (1972) *J. Biol. Chem.* **247**, 3697–3700
25. Vanderkooi, G. (1974) *Biochim. Biophys. Acta* **344**, 307–345
26. Fuller, S.D., Capaldi, R.A., & Henderson, R. (1979) *J. Mol. Biol.* **134**, 305–327
27. Kawato, S., Ikegami, A., Yoshida, S., and Orii, Y. (1980) *Biochemistry* **19**, 1598–1603
28. Kawato, S., Yoshida, S., Orii, Y., Ikegami, A., & Kinosita, K., Jr. (1981) *Biochim. Biophys. Acta* **634**, 85–92
29. Yoshida, S., Orii, Y., Kawato, S., & Ikegami, A. (1979) *J. Biochem.* **86**, 1443–1450
30. Kinosita, K., Jr., Kawato, S., Ikegami, A., Yoshida, S., & Orii, Y. (1981) *Biochim. Biophys. Acta* **647**, 7–17
31. Heyn, M.P. (1979) *FEBS Lett.* **108**, 359–364
32. Jost, P.C., Nadakavukaren, K.K., & Griffith, O.H. (1977) *Biochemistry* **16**, 3110–3114
33. Knowles, P.F., Watts, A., & Marsh, D. (1979) *Biochemistry* **18**, 4480–4487
34. Seelig, A. & Seelig, J. (1978) *Hoppe-Seyler's Z. Physiol. Chem.* **359**, 1747–1756
35. Oldfield, E., Gilmore, R., Glaser, M., Gutowsky, H.S., Hshung, J.C., Kang, S.Y., King, T.E., Meadows, M., & Rice, D. (1978) *Proc. Natl. Acad. Sci. U.S.* **75**, 4657–4660
36. Rice, D.M., Hsung, J.C., King, T.E., & Oldfield, E. (1979) *Biochemistry* **18**, 5885–5892
37. Stoeckenius, W., Lozier, R.H., & Bogomolni, R.A. (1979) *Biochim. Biophys. Acta* **505**, 215–278
38. Henderson, R. & Unwin, P.N.T. (1975) *Nature* **257**, 28–32
39. Razi Naqvi, K., Gonzalez-Rodriguez, J., Cherry, R.J., & Chapman, D. (1973) *Nature New Biol.* **245**, 249–251

40. Pasternak, C. & Shinitzky, M. (1978) in *Energetics and Structure of Halopholic Microorganisms* (Caplan, S.R. & Ginzburg, M., eds.), pp. 309–314, North-Holland Biomed. Press, New York
41. Korenstein, R. & Hess, B. (1978) *FEBS Lett.* **89**, 15–20
42. Kouyama, T., Kimura, Y., Kinosita, K., Jr., & Ikegami, A. (1981) *FEBS Lett.* **124**, 100–104
43. Kimura, Y., Ikegami, A., Ohno, K., Saigo, S., & Takeuchi, Y. (1981) *Photochem. Photobiol.* **33**, 435–439
44. Kouyama, T., Kimura, Y., Kinosita, K., Jr., & Ikegami, A. (1981) *J. Mol. Biol.* **153**, 337–359
45. Förster, T. (1948) *Ann. Phys.* **2**, 55–75
46. King, G.I., Moweny, P.C., Stoeckenius, W., Crespi, H.L., & Schoenborn, B.P. (1980) *Proc. Natl. Acad. Sci. U.S.* **77**, 4726–4730
47. King, G.I., Stoeckenius, W., Crespi, H.L., & Schoenborn, B.P. (1979) *J. Mol. Biol.* **130**, 395–404
48. Kinosita, K., Jr., Kataoka, R., Kimura, Y., Gotoh, O., & Ikegami, A. (1981) *Biochemistry* **20**, 4270–4277

# Interaction of Enveloped Virus with Cell Membranes: A Spin Label Study*

SHUN-ICHI OHNISHI AND TOYOZO MAEDA

*Department of Biophysics, Faculty of Science, Kyoto University, Kyoto 606, Japan*

Membrane fusion and splitting are involved in many of the cellular responses to outer and inner signals (*1*). The excretion of cellular products such as hormones (*e.g.*, catecholamine), chemical transmitters (acetylcholine), peptide hormones (insulin), and other secretory proteins such as zymogens, lipoproteins, albumin, and immunoglobulins is carried out by the fusion of vesicles containing those products with plasma membranes. Translocation of plasma membrane proteins from the intracellular production site, the rough endoplasmic reticulum, also involves a series of membrane fusion and splitting. The uptake of outer materials into cells often involves engulfment of these materials by a portion of the plasma membrane and splitting into small vesicles. Opsonized bacteria and other particles bind to cell surface receptors of phagocytes and are taken up into phagosomes. Excreted peptide hormones and proteins such as insulin, lipoproteins, and immunoglobulins also bind to cell surface receptors and are endocytized into coated

---

* This article is written in memory of Dr. Toyozo Maeda who died suddenly on January 8, 1981. His fine ideas on the planning of experiments and constant research activity aimed at understanding the virus-cell interaction mechanism are deeply engraved in our minds and shall never be forgotten.

vesicles (2). Phagosomes and other endocytized vesicles then fuse with lysosomes. The outer materials are digested and processed in the secondary lysosomes by a group of hydrolyzing enzymes under acidic conditions. There appear to be still other important cellular processes to be disclosed involving fusions and splittings.*1 How do these dynamic intermembrane phenomena occur? What triggers the initiation of membrane fusion and splitting?

We have been studying interactions between enveloped viruses (HVJ*2 and influenza virus) and cell membranes (human erythrocytes, ghosts, cultured cells, and liposomes) as a model system for fusion. HVJ first binds to sialoglycoproteins and sialoglycolipids on the cell surface (Fig. 1a) and then its envelope fuses with the cell membrane (envelope fusion, Fig. 1b) (4). The internal contents such as viral RNA are released into the

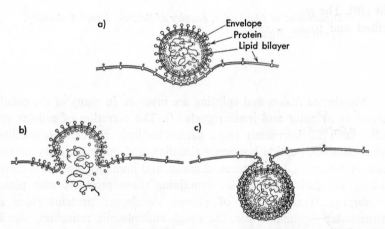

Fig. 1. Interaction between enveloped virus with cell membranes. a) Binding. b) Envelope fusion. c) Endocytosis. The virus envelope consists of proteins embedded into the lipid bilayer. The HVJ envelope contains HANA (MW 67,000) which has hemagglutinating and neuraminidase activities and F (=$F_1$ (MW 51,000)+$F_2$ (MW 15,000)) responsible for the fusion activity. The influenza virus envelope has hemagglutinin (=$HA_1$ (MW 50,000)+$HA_2$ (MW 25,000)) and neuraminidase (MW 55,000). HANA or hemagglutinin binds to sialoglycoproteins and sialoglycolipids on the target cell membranes.

---

*1 For example, nanomolar insulin caused a rapid translocation of glucose transport protein from the Golgi apparatus to the plasma membrane in fat cells (3).

*2 HVJ is an abbreviation for hemagglutinating virus of Japan, a synonym for Sendai virus.

target cell cytoplasm by this event. The cell membrane permeability transiently increases as a result of envelope fusion. This causes swelling of erythrocytes, due to water influx, and leads to hemolysis. Cell fusion also occurs (5). Influenza virus is another envelope virus, similar to HVJ,* and it also binds to sialoglycoproteins and sialoglycolipids on the cell surface. Uptake of the virus by endocytosis (called viropexis for virus uptake (6)) has been shown in morphological studies (Fig. 1c). However, whether envelope fusion also occurs is not yet clear and the mechanism of viral RNA transfer into the cytoplasm has not yet been elucidated.

We have been using spin-labeled phospholipids and a low molecular weight spin label, tempocholine, to study virus-cell interactions (see Fig. 2). The former was used to measure the transfer of phospholipid from the virus envelope to target cell membranes (7–9) and the latter to measure its release out of the preloaded virus particles upon interaction with cells (10). The purpose of this article is to describe the principles of our method and to summarize some of the results and conclusions. It is

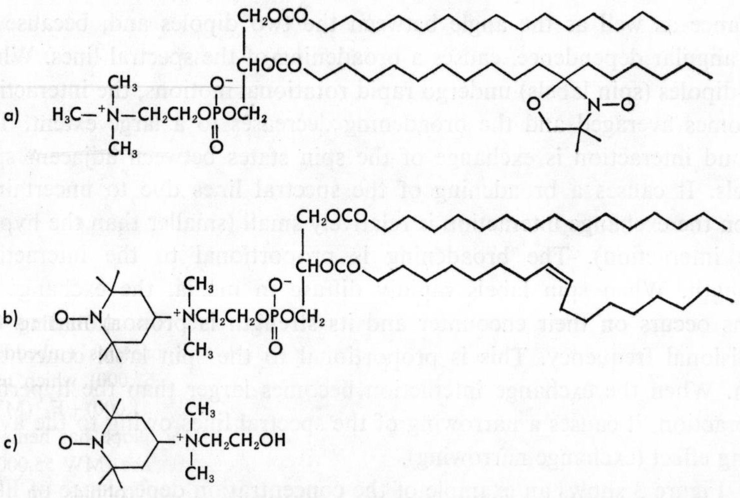

Fig. 2. Some useful spin labels for the study of virus-cell interactions. a) Phosphatidylcholine with a spin-labeled alkyl chain (PC*). b) Phosphatidylcholine analog with the spin label attached to the head group (Tempo-PC). c) A low molecular weight spin label (tempocholine).

* HVJ belongs to paramyxovirus and influenza virus to the myxovirus group.

shown that the spin labels provide a very sensitive method for envelope fusion assay. Exchange of phospholipids between the virus envelope and cell membranes is also shown to occur. Envelope fusion of influenza virus is negligibly small at neutral pH but is greatly activated under mildly acidic conditions; protonation is the trigger for rapid membrane fusion in this system.

## I.  METHODS

### 1.  Transfer of Phospholipids from Virus Envelope to Target Cell Membranes
#### 1)   Use of ESR spectra concentration dependence

The ESR spectra of spin labels in membranes and in solutions are markedly dependent on the concentration. As the concentration increases, the average distance between spin labels decreases and interaction between the unpaired spins on the labels increases. There are two types of interactions. One is the interaction between magnetic dipoles, since each spin carries a magnetic dipole moment. This is dependent on the distance as well as the angle between the two dipoles and, because of the angular dependence, causes a broadening of the spectral lines. When the dipoles (spin labels) undergo rapid rotational motions, the interaction becomes averaged and the broadening decreases to a large extent. The second interaction is exchange of the spin states between adjacent spin labels. It causes a broadening of the spectral lines due to uncertainty when the exchange interaction is relatively small (smaller than the hyperfine interaction). The broadening is proportional to the interaction strength. When spin labels rapidly diffuse in media, the exchange of spins occurs on their encounter and its strength is proportional to the collisional frequency. This is proportional to the spin label concentration. When the exchange interaction becomes larger than the hyperfine interaction, it causes a narrowing of the spectral lines owing to the averaging effect (exchange narrowing).

Figure 3 shows an example of the concentration dependence of ESR spectra; spin-labeled phosphatidylcholine (PC*) in the phosphatidylcholine bilayer membrane. Broadening of the three lines at smaller concentrations is evident. The inset graph shows a marked decrease in the ESR peak height owing to the broadening. Since the spectra are recorded as the first derivatives of ESR absorption with respect to the magnetic field,

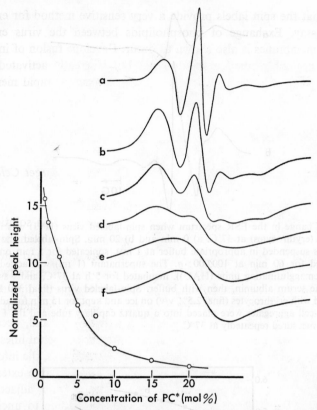

Fig. 3. Concentration dependence of the ESR spectrum of PC* in egg yolk phosphatidylcholine bilayer membranes. Concentration of PC* (mol%): a) 1, b) 15, c) 25, d) 40, e) 80, and f) 100. Spectra were measured at 3.4°C. The inset graph shows the central peak height *vs.* mol% of PC* in PC*-egg yolk phosphatidylcholine-cholesterol membranes. The molar ratio of phospholipid to cholesterol is 4:1. Spectra were measured at 23°C and integrated twice. The peak height was normalized to the same double-integrated area.

the peak height is approximately proportional to the inverse square of the line width. In the following assay, we simply follow the change (increase) in ESR peak height due to dilution of spin labels.* As the concentration increases further, the three lines are broadened to a greater

---

* Of course we can make a plot of the line width or the line width parameter *vs.* the concentration and use the parameter for the analysis. This was used for the analysis of Ca²⁺-induced phase separations (*11*), but is not suitable for the present system.

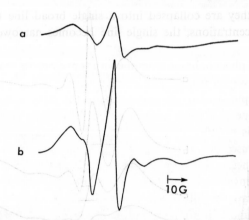

Fig. 4. Change in the ESR spectrum when spin-labeled virus (HVJ) was incubated with cells (erythrocytes) at 37°C. a) 0 min and b) 20 min. Spin-labeled phospholipid (PC*) was suspended in appropriate buffer at 1 mM, sonicated for 5 min on ice, and centrifuged for 60 min at $100,000 \times g$. The supernatant (1 ml) was added to HVJ (20,000 hemagglutinating units (HAU)), incubated for 5 h at 37°C, and washed with 1% bovine serum albumin, then with buffer. Spin-labeled virus (final 300 HAU/ml) was mixed with erythrocytes (final 2.5% v/v) on ice and kept for 15 min for adsorption. The virus-cell aggregates were placed into a quartz capillary tube and the ESR spectrum was measured repeatedly at 37°C.

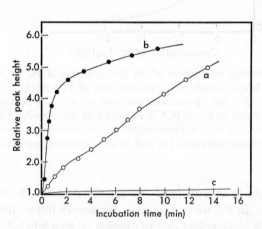

Fig. 5. Increase in the ESR peak height due to transfer of PC* from the virus envelope to cell membranes at 37°C. a) HVJ-erythrocytes at neutral pH, b) influenza virus-erythrocytes at pH 5.2, and c) influenza virus-erythrocytes and trypsinized HVJ-erythrocytes at neutral pH. The ordinate plots the central peak height at time $t$ divided by that at time 0. Spin labeling of the virus and ESR measurements of the virus-cell aggregates are as described in Fig. 4.

extent so that they are collapsed into a single broad line (Fig. 3e). At still higher concentrations, the single line becomes narrowed (exchange narrowing).

Transfer of phospholipids from the virus envelope to the target membrane can be assayed using the spectra concentration dependence. For this purpose, spin-labeled phospholipids (e.g., PC*) are incorporated into the envelope at a high concentration (10–15 mol%), so that the ESR spectrum is broadened by the spin-spin interactions (e.g., Fig. 4a). When these viruses are adsorbed onto the cell surface and incubated, the ESR peak height will increase when the spin-labeled phospholipids move from the envelope to the target cell membrane, since the transferred phospholipids rapidly diffuse away in the cell membrane and become diluted. An example of the spectral change is shown in Fig. 4 and an example of the peak height increase in Fig. 5. The peak height increase approaches a 6-fold level. This is reasonable judged from the result for

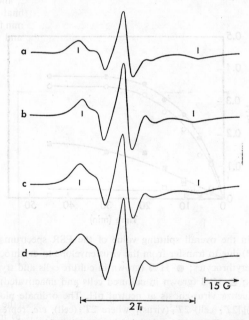

Fig. 6. ESR spectrum of spin-labeled phospholipid (PC*) incorporated at a small concentration into the membranes of erythrocytes (a), ghosts (b), HVJ (c), and influenza virus (d) at neutral pH at 23°C. The overall splitting value ($=2T_{\parallel}$) is 52 G (a), 49 G (b), 51 G (c), and 50 G (d).

the model membrane (Fig. 3), which shows a 6-fold increase when PC* at a concentration of 12% is diluted to 3%.

A similar assay can be done by utilizing any physical phenomena that show concentration dependence. Quenching of fluorescence is one such example. Wyke *et al.* (*12*) used fluorescence enhancement when anthroylstearate is transferred from the HVJ envelope to cell membranes.

*2) Use of membrane fluidity change*

The ESR spectrum of spin-labeled phospholipids incorporated in membranes gives information on the fluidity of membranes surrounding them. Figure 6 shows some examples for PC* incorporated in various membranes at a small concentration to avoid the spin-spin broadening. The overall splitting value $2T_\parallel$ is used as a measure of the fluidity*; a larger splitting value corresponds to less fluidity. The fluidity of the erythrocyte membrane is different from that of the ghost membrane prepared by various treatments (osmotic (*14*), virus (*15*), antibody, and

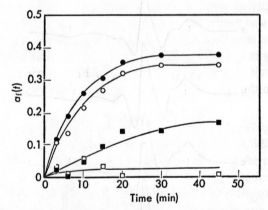

Fig. 7. Change in the overall splitting value of the ESR spectrum of spin-labeled phospholipid (PC*) due to transfer from the virus envelope to erythrocyte membranes at 37°C. ○ HVJ-erythrocytes; ● HVJ (grown in culture cells and trypsin activated)-sheep erythrocytes; ■ HVJ (grown in cultured cells and unactivated)-sheep erythrocytes; and □ influenza virus-ghosts at neutral pH. The ordinate plots $a_f(t)$ defined as $[2T_\parallel(t)-2T_\parallel(0)]/[2T_\parallel(\text{cell})-2T_\parallel(\text{virus})]$ where $2T_\parallel(\text{cell})$, *etc.* represent the overall splitting value for the cell, *etc.* The spin-label has the nitroxide moiety at the 7th position. (Taken from Lyles and Landsberger (*17*))

---

* One can also use the order parameter (see Ref. *13*).

complement (*16*)). The overall splitting value for erythrocytes is 52 G at 23°C to be compared with 49 G for ghosts. Erythrocyte membranes are therefore more rigid than ghost membranes. PC* in the HVJ envelope gives 52 G, indicating that the virus envelope is more rigid than ghost membranes.

The difference in the fluidity between the virus envelope and ghost membranes was utilized by Lyles and Landsberger (*17*) to assay the transfer of phospholipids from the envelope to erythrocyte membranes. The overall splitting value will decrease when the spin-labeled phospholipids are transferred. An example of the change in the splitting value is shown in Fig. 7 where the ordinate plots $a_f(t)$ as defined in the legend. An increase in this parameter corresponds to a decrease in the splitting value. Similarly, the fluidity difference between erythrocytes and ghosts was used to assay virus-induced hemolysis. For this purpose, unlabeled virus was adsorbed onto spin-labeled erythrocytes and incubated. A decrease in the overall splitting value followed.

Spin-labeled phospholipids should be able to give information on possible fluidity changes in the target cell membranes caused by virus. When HVJ was incubated with spin-labeled erythrocytes, the overall splitting value does decrease (from 52 to 49 G). However, this increase in the fluidity may not represent the effect of the virus itself but merely reflect the result of virus-induced hemolysis. Further careful studies need to be done on this subject.

## 2. Release of Tempocholine from Preloaded Virus Particles (10)
### 1) Use of ESR peak height increase on dilution

Virus particles can be loaded with low molecular weight spin labels simply by incubating them in media containing the spin labels and washing by centrifugation. Tempocholine can be loaded to higher concentrations (20–30 mM), probably because of its positive charge, so that the ESR spectrum is broadened by the spin-exchange interaction (see Fig. 8Aa). When these virus particles are adsorbed onto the cell surface and incubated, the ESR peak height will increase when the inside tempocholine leaks out of the virus. When envelope fusion occurs, the internal tempocholine will be rapidly released into the target cell cytoplasm, diluted, and its ESR peak height will increase. An example of the spectral change is shown in Fig. 8A and an example of the peak height increase in Fig. 8B. The peak height goes up to a 5-fold increase level. A model

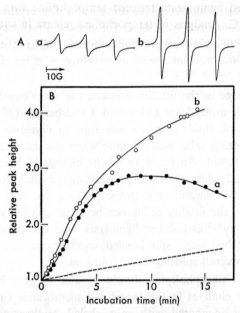

Fig. 8.   Assay of tempocholine release from the preloaded virus particles. A) Change in the ESR spectrum of tempocholine-loaded HVJ on incubation with ghosts at 37°C: a) 0 min and b) 15 min. B) Increase in the ESR peak height *vs.* incubation time: a) HVJ-erythrocytes and b) HVJ-ghosts. The dotted line shows a level for spontaneous leakage from the loaded virus in the absence of cells. The release from trypsinized HVJ on incubation with erythrocytes approximately followed the dotted curve. HVJ ($2.5 \times 10^4$ HAU) was loaded by incubating in 0.5 ml of 140 mM tempocholine at 30°C for 4 h, washed 3 times with phosphate-buffered saline, and resuspended in 1 ml buffer. The loaded virus (final 300 HAU/ml) was mixed with erythrocytes (final 2.5% v/v) at 0°C, kept for 10 min for adsorption, and centrifuged for 3 min at $350 \times g$. The pellet was placed into a quartz capillary tube and the ESR spectrum was recorded repeatedly at 37°C.

experiment shows that a 5-fold increase occurs when 20 mM tempocholine aqueous solution was diluted to less than 1 mM.

Cell cytoplasms often contain reducing agents. When incubated with such cells, the spin-label nitroxide group will be reduced upon contact with the cytoplasm and lose its ESR signal. Curve a in Fig. 8B shows an example for the spin-label reduction when tempocholine-loaded virus was incubated with erythrocytes. The peak height increases in the initial stage due to dilution but decreases in the later stage due to the reduction. Erythrocyte cytoplasm does contain reducing agents;

freshly prepared hemolyzate reduced tempocholine with a half time of 6.5 min at 37°C. Analysis of tempocholine release in such systems may be somewhat complicated because of the coexistence with the reduction.

*2) Use of reduction of spin labels accessible to external reagents*

There are some reagents that reduce the spin label nitroxide to hydroxylamine and reagents that reoxidize the hydroxylamine to nitro-

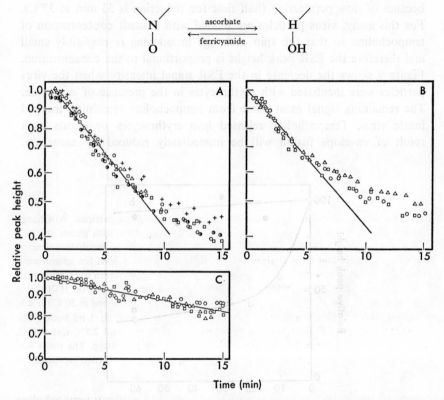

Fig. 9. Assay of tempocholine remaining trapped inside the virus particles on incubation with cells. A) HVJ-erythrocytes and B) HVJ-ghosts. The virus was loaded with tempocholine in the same way as described in Fig. 8 except for using 1 mM tempocholine instead of 140 mM. The loaded virus at various concentrations was mixed with erythrocytes or ghosts (final 2.5% v/v) and incubated at 37°C in the presence of 1 mM ascorbate. The ESR peak height was measured repeatedly. Virus concentration (final): + 104; □ 208; ○ 416; △ 832; and ◐ 1,234 HAU/ml. C) Tempocholine remaining trapped inside trypsinized HVJ upon incubation with erythrocytes at 37°C. HVJ at various final concentrations was coadsorbed: ○ 0; □ 312; and △ 1,248 HAU/ml.

xide. Ascorbate and ferricyanide are such examples. If these reagents are not permeable to membranes, we could discriminate spin labels located outside the membranes from those inside. For example, we could quantitate tempocholine remaining inside virus particles by using ascorbate. Ascorbate immediately reduces tempocholine in water (in less than 1 min) but takes time to reduce tempocholine inside a virus because of slow permeation (half time for reduction is 53 min at 37°C). For this assay, virus particles are loaded with a small concentration of tempocholine so that the spin exchange broadening is negligibly small and therefore the ESR peak height is proportional to the concentration. Figure 9 shows the decrease in the ESR signal intensity when the virus particles were incubated with erythrocytes in the presence of ascorbate. The remaining signal must come from tempocholine remaining trapped inside virus. Tempocholine released into erythrocytes or ghosts as a result of envelope fusion will be immediately reduced by ascorbate.

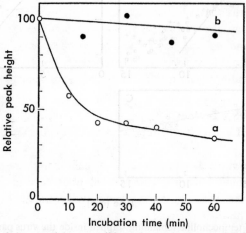

Fig. 10. Assay of virus particles accessible to the outer reagent. a) Influenza virus labeled with Tempo-PC was pretreated with 10 mM ascorbate at 4°C and washed. The virus (final 3,000 HAU/ml) was mixed with Madin-Darby Canine Kidney cells ($5 \times 10^8$/ml), kept for 60 min on ice for adsorption, and washed. The virus-cell aggregates were suspended in MEM-HEPES at pH 7.4 and incubated at 37°C. An aliquot was withdrawn at the indicated time and centrifuged at 4°C. Ferricyanide (10 μl, 10 mM in the buffer) was added to the pellet and the ESR spectrum was measured at 23°C. b) A control experiment where the cells were pretreated with 2.5% glutaraldehyde for 30 min at room temperature.

Ascorbate may be able to enter the cells upon lysis or due to the increased membrane permeability brought about by envelope fusion.

### 3. Assay of Endocytized Virus Particles

Phospholipids with the spin label attached to the head group may be used to distinguish virus particles bound outside the cell surface from those taken up inside the cells. Tempo-PC (see Fig. 1) incorporated into virus envelopes was reduced instantaneously on addition of ascorbate to the medium and reoxidized immediately by ferricyanide. The spin-labeled virus and also ascorbate-treated virus were used for the assay of endocytosis. An example is shown in Fig. 10 where the ascorbate-treated virus was incubated with MDCK cells at 37°C. The ESR spectrum was measured after addition of ferricyanide. The restored signal comes from spin labels of the virus particles bound on cell surface and also from those of virus fused with cell membranes. The unrestored signal can be taken as due to the endocytized virus particles. The half time for endocytosis is read as 7 min from the curve (a). Similar result was obtained when the spin-labeled virus (without ascorbate pretreatment) was incubated with cells. In this assay, the ESR signal appears to come from the virus particles remaining bound on cell surface and those fused with cell membranes since it disappeared on addition of ascorbate. In other words, the spin labels on the endocytized virus particles had been somehow reduced.

## II. ANALYSIS OF PHOSPHOLIPID TRANSFER AND TEMPO-CHOLINE RELEASE BASED ON ENVELOPE FUSION

When the viral envelope fuses with cell membranes, the envelope proteins and lipids, including spin-labeled phospholipids, would rapidly diffuse out of the fused site and intermix well with the target cell membrane lipids. The ESR peak height of the spin labels will increase owing to a weakening of the spin-spin interactions. The intermixing occurs very fast, much faster than envelope fusion. Suppose a virus with a diameter of 0.1 $\mu$m fused with the cell membrane and diffused laterally with a diffusion constant of $10^{-8}$ cm$^2$/s, and then the concentration of spin-labeled phospholipids in the fused site will decrease to 1/5 in 0.3 s.

When virus preloaded with a high concentration of tempocholine is used, the ESR peak height will also increase rapidly upon envelope fusion because of dilution into the cytoplasm. When lightly-loaded virus is used in the presence of ascorbate, the ESR signal will decay rapidly upon envelope fusion owing to the nitroxide reduction.

Suppose envelope fusion occurs with a rate constant $k$, then the number of envelope-fused virus particles at a time $t$ is given by

$$n/N_0 = F = 1 - e^{-kt} \qquad (1)$$

where $N_0$ is the number of virus particles bound on the cell surface at time 0. $F$ is the fraction of fused virus. The spin labels diluted in the cell membrane (for PC*) or in the cytoplasm (for tempocholine) give an $f$-times larger peak height than that when they were in the envelope or inside the virus particle. Then the peak height at time $t$, $I_t$, divided by that at time 0, $I_0$, is given by

$$h = I_t/I_0 = f - (f-1)e^{-kt}. \qquad (2)$$

In the initial stage of the reaction where $(kt/2) \ll 1$

$$h \doteqdot 1 + (f-1)kt. \qquad (3)$$

Therefore we can obtain $(f-1)k$ from the initial slope and $f$ from the final level of the peak height increase curve. Combination of Eqs. (1) and (2) leads to

$$(f-h)/(f-1) = e^{-kt} = B \qquad (4)$$

where $B = 1 - (n/N_0)$ is the fraction of virus remaining bound on the cell.

The ESR signal decay for lightly-loaded virus in the presence of ascorbate can be analyzed in the same way. Here the remaining signal directly represents $B$ and a plot of its logarithm against time should give a straight line. This was actually observed in the initial stage of the reaction (Fig. 9). The envelope fusion rate constant $k$ is obtained from the slope as 0.078 min$^{-1}$ for erythrocytes and 0.086 min$^{-1}$ for ghosts. In the later stage, the data points deviate towards larger $B$ values, $i.e.$, less efficient envelope fusion. This means that not all virus particles fuse with cells in the same way. Some populations may fuse more slowly. Some may be eluted from the cell surface during incubation due to the sialidase activity of the virus.

a)  *Phospholipid transfer*  From the initial slope of the curve (Fig. 5a), $(f-1)k = 0.5$ min$^{-1}$ is obtained. The $f$ value should be obtainable from the saturation level of the curve, provided that the transfer is solely due

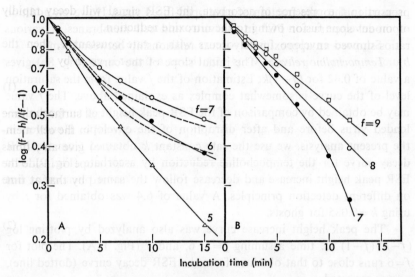

Fig. 11. Analysis of the ESR peak height increase curve for tempocholine release (Fig. 8B, curve b) (A) and for PC* transfer (Fig. 5, curve a) (B) based on envelope fusion (Eq. (4) in the text). $h$, the peak height at time $t$; $f$, the peak height increase factor upon dilution.

to envelope fusion until the end of the incubation and all virus particles fuse at the same rate. However, this is not the case as inferred from Fig. 9A and B. An alternative method for the estimation of $f$ is to compare the ESR peak height of PC* incorporated in the cell membrane at a small concentration with that of PC* incorporated in the virus envelope, both normalized to the same double-integrated area. The ratio is equal to $f$ and was obtained to be 6–7 (K. Kuroda, personal communication). The $f$ value gives a $k$ value of 0.1–0.08 $min^{-1}$ at 37°C.

The data can also be analyzed by plotting $\log (f-h)/(f-1)$ vs. incubation time. Figure 11B shows such a plot for three different $f$ values. The rate constant $k$ is calculated to be 0.093 for $f=6$ and 0.076 for $f=7$ from the initial slope. In the later stage, the data points tend to deviate from the straight line towards lower $B$ values, i.e., more efficient transfer.

In the above assays we measure the central peak height for the estimation of $I_t$ and $I_0$. Since the peak position for PC* in the virus envelope is somewhat different from that for cell membranes, owing to the broadening, the peak height increase $h$ thus obtained may not be rigorously proportional to the fraction of enveloped-fused virus. However, addition of model spectra for PC* in the virus and cell membranes at various ratios showed an approximately linear relationship between them.

b) *Tempocholine release* The initial slope of the curve (Fig. 8b) gives a value of 0.42 for $(f-1)k$. Estimation of the $f$ value from the saturation level of the curve is somewhat complex as explained above. The $f$ value may be obtained by comparison of the ESR peak height of tempocholine-loaded virus before and after disruption of the envelope. However, in the present analysis, we use the rate constant $k$ obtained from the ESR decay curve for the tempocholine reduction by ascorbate (Fig. 9). The ESR peak height increase and decrease follow the same phenomena but on different detection principles. A value of 6.4 was obtained for $f$ by using $k=0.085$ for ghosts.

The peak height increase curve was also analyzed by plotting log $(f-h)/(f-1)$ *vs.* time assuming $f=5$, 6, and 7 (Fig. 11A). The plot for $f=6$ runs close to that obtained from the ESR decay curve (dotted line), as was expected.

The above analyses give the rate constant $k$ to be 0.08–0.09 min$^{-1}$ at 37°C for envelope fusion of HVJ with erythrocytes.* The half time $t_{1/2}$ for the envelope fusion $(t_{1/2}=\ln 2/k)$ is therefore 8.7–7.7 min. The envelope fusion rate was independent of the virus dose and virtually the same with erythrocytes as with ghosts. The assay of envelope fusion by the peak height increase has the advantage that it measures the rate constant multiplied by $(f-1)$; it sensitively detects the reaction at a magnification of 5–6.

Lyles and Landsberger (*17*) analyzed the change in the overall splitting value when HVJ was incubated with erythrocytes (Fig. 7). The time course of $a_f(t)$ followed

$$a_f(t)=a_f(\infty)(1-e^{-kt}) \tag{5}$$

where $a_f(\infty)$ is the value of $a_f$ at infinite time. The envelope fusion rate constant $k$ was obtained to be 0.1 min$^{-1}$ at 37°C. These authors also observed that the fusion rate was the same between erythrocytes and ghosts and was independent of the virus dose. Virus-induced hemolysis

---

* When earlier phase of the phospholipid transfer was measured more carefully and the peak height increase curve was analyzed, the result revealed another much faster component with the rate constant of the order of 0.01–0.02 s$^{-1}$ (K. Kuroda, personal communication).

took place at a rate similar to that of envelope fusion, suggesting that envelope fusion is the rate-limiting step in hemolysis.

Phospholipid transfer and tempocholine release occur only with HVJ containing active F protein. When trypsinized HVJ which has cleaved F protein, or HVJ grown in cultured cells containing the precursor form of F protein, was incubated with erythrocytes, those reactions occurred only to negligibly small extents (Figs. 5c, 8, and 9C). Active F protein is required for envelope fusion. Envelope fusion of influenza virus was negligibly small at neutral pH (Fig. 5c).

## III. PHOSPHOLIPID TRANSFER *VIA* EXCHANGE BETWEEN VIRUS ENVELOPE AND CELL MEMBRANE

Phospholipid transfer from trypsinized HVJ to erythrocytes was very small because of the inactive F protein. However, when intact HVJ was coadsorbed onto erythrocytes, the transfer from the trypsinized virus was greatly enhanced (8, 9). The enhancement increased with the amount of coadsorbed HVJ (Fig. 12). Tempocholine release from trypsinized HVJ upon incubation with erythrocytes was also limited to a low level in line with the phospholipid transfer. However, a markedly different feature from the latter is that the coadsorbed HVJ did not enhance tempocholine release (Fig. 9C). These results indicate that the coadsorbed HVJ does not cause envelope fusion of trypsinized HVJ with cell membranes but enhances phospholipid transfer from trypsinized HVJ. The enhancement must therefore be due to exchange of phospholipids between the trypsinized virus envelope and the erythrocyte membrane. When the target cells are ghosts, phospholipid transfer from trypsinized HVJ was not enhanced at all by the coadsorbed intact HVJ.

In the above experiments, the coadsorbed HVJ fused with erythrocyte membranes. The envelope fusion must somehow modify the target membrane and enhance the phospholipid exchange through the modification. We have proposed that the viral F proteins implanted and diffused into the target membranes may enhance phospholipid exchange (9). In order to explain lack of the enhancement in ghosts by this mechanism, we must assume only limited diffusions of the implanted viral proteins in ghost membranes. In our more recent paper (10), we propose that the cell swelling caused as a result of envelope fusion may also be responsible for the enhancement of phospholipid exchange. This can

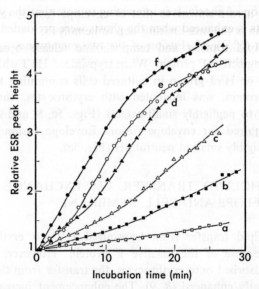

Fig. 12. Transfer of phospholipids from trypsinized HVJ to erythrocytes activated by intact HVJ. PC*-labeled trypsinized HVJ (final 260 HAU/ml) and intact HVJ at various concentrations were mixed with erythrocytes (final 2.5% v/v) and incubated at 37°C. The ESR peak height was measured repeatedly. HVJ final concentration: a) 0, b) 40, c) 160, d) 320, e) 640, and f) 1,280 HAU/ml.

explain the difference between erythrocytes and ghosts in the enhancement very well, since essentially no swelling would occur for ghosts because of the absence of intracellular proteins.

Phospholipid transfer from the influenza virus envelope to erythrocytes was negligibly small at neutral pH but when HVJ was coadsorbed it was greatly enhanced, probably by the same mechanism as above. Phospholipid transfer from HVJ to erythrocytes seems to be autocatalytically enhanced by a similar mechanism of the one above. The positive deviation in the phospholipid transfer shown in Fig. 11B suggests such an enhancement. The transfer from HVJ to erythrocytes was nearly the same as that to ghosts in the initial stage but became larger in the later stage (data not shown).

The importance of osmotic swelling in virus-induced cell fusions was emphasized by Knutton and Bächi (19); the swelling is the driving force for the polyerythrocyte formation. This is in line with the experimental observation that HVJ causes efficient fusion of erythrocytes but only slight fusion of ghosts. It is interesting to note that the virus-induced fusion of ghosts is enhanced when the ghosts were preloaded with bovine serum albumin (5%))(20). Fusion of the two membranes appears to be enhanced when they are expanded by swelling. The enhancement of phospholipid transfer may therefore be related to cell fusions.

## IV.   ENVELOPE FUSION OF INFLUENZA VIRUS IS ACTIVATED UNDER MILDLY ACIDIC CONDITIONS

Phospholipid transfer from the influenza virus envelope to erythrocyte membranes was negligibly small at neutral pH but greatly accelerated in mildly acidic pH (Fig. 5b). The maximum enhancement occurred at pH 5.2 as shown in Fig. 13A (21). This provides strong evidence for activation of envelope fusion being triggered by protonation. The acid-activated transfer was much faster than that from HVJ to erythrocytes (Fig. 5a). The envelope fusion rate constant $k$ was estimated from the initial slope to be 0.8 min$^{-1}$ for $f=7$ at pH 5.2. This value is about one order of magnitude larger than that for HVJ-erythrocytes at neutral pH.

Hemolysis activity of influenza virus at neutral pH was very low but was greatly enhanced under acidic conditions (21), in line with the activation of envelope fusion. Its pH dependence was virtually the same as that for phospholipid transfer (Fig. 13B). The acid activation also caused efficient fusion of erythrocytes. Influenza virus enhanced the phospholipid transfer from trypsinized HVJ to erythrocytes in acidic conditions, though no enhancement was seen at neutral pH. All these results indicate that what HVJ does on erythrocytes in neutral pH is done by influenza virus in acidic conditions.

The marked difference between influenza virus and HVJ in their pH characteristics can be ascribed to the envelope constituents, especially proteins, but not to the target membranes. For example, phospholipid transfer from influenza virus to liposomes showed essentially the same pH profile as that to erythrocytes as well as to other cultured cells (22). Phospholipid transfer from HVJ to liposomes did not occur at neutral as well as acid pH values. F proteins ($=F_1+F_2$) in HVJ and hemagglutinin ($=HA_1+HA_2$) in influenza virus are responsible for the fusion activity. The amino termini of these proteins contain a considerable number of hydrophobic amino acid residues and are thought to be important for the interaction with the target membranes (23). The $F_1$ terminus

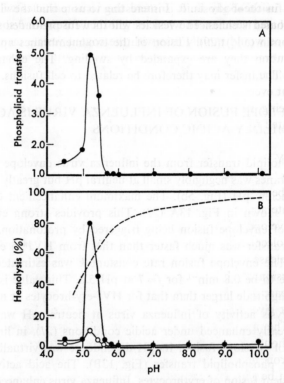

Fig. 13. pH dependence of phospholipid transfer from the influenza virus envelope to erythrocyte membranes (A) and of the virus-induced hemolysis (B). A) PC*-labeled virus (final 33 μg protein/ml) was mixed with erythrocytes (final 2.5% v/v) and incubated at 37°C. The ordinate plots the ratio of the ESR peak height at 5 min to that at 0 min. B) Influenza virus (15 μg protein/ml) was mixed with erythrocytes (2.5% v/v) at various pH values and incubated for 30 min at 37°C. ○ is for the virus grown in cultured cells and unactivated. The dotted line shows pH dependence of HVJ-induced hemolysis.

consists of 15 hydrophobic residues and the $HA_2$ terminus contains 21 hydrophobic residues interrupted by a few acidic residues, glutamate and aspartate (24). We have proposed that protonation of the acidic residues would cause conformational change in the $HA_2$ amino terminus which enhances interaction with the target cell membranes (22).

When influenza virus invades animals, it will bind to the target cell surface but may not be able to fuse with the cell membrane because of neutral pH in the body fluid. Instead the virus particles will be endo-cytized in coated vesicles. The vesicles will then fuse with lysosomes. The viral envelope would rapidly fuse with the coated vesicle membrane (ex-plasma membrane) because of the low pH (4.5–4.9) inside the secondary lysosomes. The virus genome will be transferred into the cytoplasm through that event. Experimental evidence for this infectious entry mech-anism is being accumulated (*18*). The same infection mechanism has been presented for Semliki Forest virus, a togavirus (*25*).

## SUMMARY

It has been shown that spin labels can be used successfully to study virus-cell interaction mechanisms. They provide a unique means for the assay of envelope fusion. The method is based on the transfer of phos-pholipid molecules from the virus envelope to target cell membranes and also release of the internal contents of virus particles. It detects envelope fusion at a high magnification since it utilizes the ESR peak height increase accompanied by the transfer or release at a factor of 6-to 7-fold. It gives quantitative data on envelope fusion such as the rate constant. The method measures the average value of some $10^9$–$10^{10}$ viruses, although at the expense of the individuality of the virus and cells. Electron microscopy can observe envelope fusion more directly but may not be suitable for quantitative measurements.

Analysis of the spin label data in the initial stage of virus-cell inter-actions gives a value of 0.08–0.09 min$^{-1}$ at 37°C for the envelope fusion rate constant of HVJ with erythrocytes. The rate was not much different between erythrocytes and ghosts and was independent of the virus dose. The envelope fusion of influenza virus was negligibly small at neutral pH but greatly activated in mildly acidic conditions (pH 5.2). The acti-vated envelope fusion was much faster than that of HVJ with erythro-cytes at neutal pH, with the rate constant being one order of magnitude larger. An infectious cell entry mechanism of influenza virus is proposed based on this finding.

In later stages of virus-cell interactions, some enhancement effect appears when the target cells were erythrocytes. Exchange of phospho-lipids between the HVJ envelope and erythrocyte membrane was accel-erated. This enhancement was not observed, however, when the target cells were ghosts and related to swelling of cells caused as a result of envelope fusion.

Spin-labeled phospholipids with the nitroxide moiety attached to the head group can be used to assay number of endocytized virus particles.

*Acknowledgment*
    We thank Akira Asano, Yoshio Okada, Sakuji Toyama, Kazumichi Kuroda, Kazunori Kawasaki, and Akihiko Yoshimura for their cooperation and contributions.

# REFERENCES

1. A general reference to this subject: Poste, G. & Nicolson, G.L. (eds.) *Cell Surface Reviews* Vol. 4 *Membrane Assembly and Turnover* (1977) and Vol. 5 *Membrane Fusion* (1977) North-Holland, Amsterdam, New York, and Oxford
2. Goldstein, J.L., Anderson, R.G.W., & Brown, M.S. (1979) *Nature* **279**, 679–685
3. Suzuki, K. & Kono, T. (1980) *Proc. Natl. Acad. Sci. U.S.* **77**, 2542–2545
4. Poste, G. & Nicolson, G.L. (eds.) (1978) *Cell Surface Reviews*, Vol. 2, *Virus Infection and Cell Surface*, North-Holland, Amsterdam, New York, and Oxford
5. Okada, Y. (1958) *Biken J.* **1**, 103–110
6. Fazekas de St. Groth, S. (1948) *Nature* **162**, 294–295
7. Maeda, T., Asano, A., Ohki, K., Okada, Y., & Ohnishi, S. (1975) *Biochemistry* **14**, 3736–3741
8. Maeda, T., Asano, A., Okada, Y., & Ohnishi, S. (1977) *J. Virol.* **21**, 232–241
9. Kuroda, K., Maeda, T., & Ohnishi, S. (1980) *Proc. Natl. Acad. Sci. U.S.* **77**, 804–807
10. Maeda, T., Kuroda, K., Toyama, S., & Ohnishi, S. (1981) *Biochemistry* **20**, 5340–5345
11. Ohnishi, S. & Tokutomi, S. (1981) *Biol. Magn. Reson.* **3**, 121–153
12. Wyke, A.M., Impraim, C.C., Knutton, S., & Pasternak, C.A. (1980) *Biochem. J.* **190**, 625–638
13. Berliner, L.J. (ed.) (1976) *Spin Labeling Theory and Applications*, Academic Press, New York, San Francisco, and London
14. Tanaka, K. & Ohnishi, S. (1976) *Biochim. Biophys. Acta* **426**, 218–231
15. Lyles, D.S. & Landsberger, F.R. (1977) *Proc. Natl. Acad. Sci. U.S.* **74**, 1918–1922
16. Nakamura, M., Ohnishi, S., Kitamura, H., & Inai, S. (1976) *Biochemistry* **15**, 4838–4843
17. Lyles, D.S. & Landsberger, F.R. (1979) *Biochemistry* **18**, 5088–5095
18. Yoshimura, A., Kuroda, K., Kawasaki, K., Yamashina, S., Maeda, T., & Ohnishi, S. (1982) *J. Virol.* **43**, in press
19. Knutton, S. & Bächi, T. (1980) *J. Cell Sci.* **42**, 153–167

20. Sekiguchi, K. & Asano, A. (1978) *Proc. Natl. Acad. Sci. U.S.* **75**, 1740–1744
21. Maeda, T. & Ohnishi, S. (1980) *FEBS Lett.* **122**, 283–287
22. Maeda, T., Kawasaki, K., & Ohnishi, S. (1981) *Proc. Natl. Acad. Sci. U.S.* **78**, 4133–4137
23. Gething, M.-J., White, J.M., & Waterfield, M.D. (1978) *Proc. Natl. Acad. Sci. U.S.* **75**, 2737–2740
24. Gething, M.-J., Bye, J., Skehel, J., & Waterfield, M. (1980) *Nature* **287**, 301–306
25. Helenius, A., Kartenbeck, J., Simons, K., & Fries, E. (1980) *J. Cell Biol.* **84**, 404–420

# Mechanism of HVJ-induced Cell Fusion[*1]

AKIRA ASANO[*2] AND KIMIKO ASANO[*3]

*Institute for Protein Research, Osaka University, Osaka 565, Japan*[*2]
*and College of Medical Technology, Kyoto University,*
*Kyoto 606, Japan*[*3]

The fusion of biological membranes seems to be an important and fundamental event among the cellular functions. At the subcellular level, membrane fusion such as pinocytosis, phagocytosis, and phagosome-lysosome fusion, as well as exocytotic processes is widely observed in most cells (*1*). Fusion between plasma membranes is known to occur in myoblast formation, fertilization, and pathological syncytium formation.

Although the mechanisms of these membrane fusion reactions at the cellular and subcellular levels may not be completely the same, there will be a common mechanism in the fusion reaction, since the basic structure of biological membranes is lipid bilayers, and the principle to construct this structure is hydrophobic interaction, which exhibits no specificity among membrane components. Thus, close contact of two membranes and some perturbation of the bilayer structure may lead to

---

[*1] HVJ is an abbreviation for hemagglutinating virus of Japan, a synonym for Sendai virus.

[*2] Present address: Cancer Research Institute, Sapporo Medical College, Sapporo 060, Japan.

a continuation of the two membranes, *i.e.*, membrane fusion. But close apposition of two biological membranes does not always result in fusion of these membranes; therefore, there must also be mechanisms which suppress the occurrence of membrane fusion between biological membranes.

The HVJ-induced cell fusion reaction discovered by Okada (*2*) has served as a good experimental model for studies of membrane fusion, because of this system rapid and extensive membrane fusion can be attained under easily controllable conditions (*3*). The cell fusion reaction seems to be regulated by viral and cellular factors. For example, limited proteolytic processing of the virus particles is required for fusion-inducing (fusogenic) activity of the virion (*4*), and a high ATP level within cells is a prerequisite for extensive fusion of Ehrlich ascites tumor cells (*5, 6*).

In this article, we will describe 1) the reaction steps of HVJ-induced fusion reaction; 2) the structure-function relationship of viral factors which are required for the fusion reaction; and 3) cellular factors which control cell fusion.

## I.  VIRUS AND CELLS

HVJ belongs to the paramyxovirus group, a negative-strand RNA virus, and is enveloped by a membrane which mainly consists of lipid

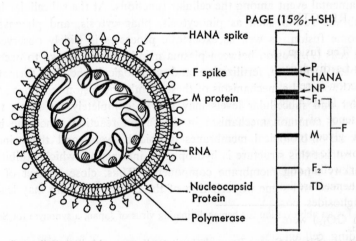

Fig. 1.   Model of HVJ and sodium dodecyl sulfate-polyacrylamide gel electrophoresis (SDS-PAGE) pattern of HVJ proteins.

bilayers. As shown in Fig. 1, the HVJ envelope contains two kinds of glycoproteins, the larger one (HANA) has hemagglutinating and neuraminidase (NA) activities (7, 8), and the smaller one, designated F, is required for membrane fusion (4), and is therefore also required for the other biological activities of the virus, such as hemolysis, cell fusion, and infection. These glycoproteins project more than 10 nm from the surface of the virion and are thus called spikes. These spikes densely cover the surface of the virus particles. The other viral proteins do not participate in the fusion reaction as shown below.

Human erythrocytes are extensively used for the study of membrane fusion reactions, because they seem to have several advantages over nucleated cells. One advantage is that erythrocytes are devoid of intracellular organelles, and their plasma membrane can be prepared easily by simple hypotonic hemolysis. Furthermore, the erythrocyte membrane is one of the most extensively characterized biomembranes. As we reported previously (9, 10), HVJ-induced fusion of human erythrocyte ghosts can be observed successfully provided that bovine serum albumin is sequestered within the ghosts at concentrations of around 5% (w/w). Therefore, the reaction sequences of cell fusion will be explained using this system. Those with nucleated cells are essentially the same as ghost fusion in most respects. Some differences among these systems will be described later in this article.

## II.  REACTION STEPS OF VIRUS-INDUCED FUSION REACTION

The fusion reaction of human erythrocyte ghosts induced by HVJ can be divided into several stages as depicted in Fig. 2. Stage I of the reaction is cell agglutination. This stage is dependent on the presence of active HANA glycoprotein on the surface of the virus particles, and the presence of virus receptors on the plasma membrane. Both glycolipids and glycoproteins which contain terminal acetylneuraminic acid are known to serve as binding sites for the HANA spikes. Recent studies by Markwell et al. (11, 12) showed that several kinds of specific carbohydrate sequences are important for the receptor function of HVJ. For example, gangliosides containing NeuAc$\alpha$2,3Gal$\beta$1,3GalNAc (i.e., GD$_{1a}$, GT$_{1b}$, and GQ$_{1b}$) were active as receptors of the virus. As the result of such binding, cell agglutination occurs, and this reaction can proceed at 4°C.

Stage II is viral envelope-cell membrane fusion. In this step, the

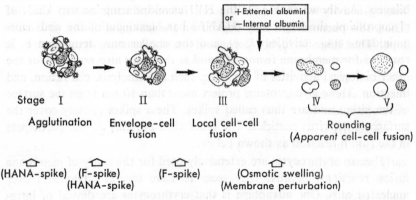

Fig. 2. Schematic representation of HVJ-induced cell fusion reaction. Stage III, local cell-cell fusion, is defined as membrane fusion in which cell membranes and the cytoplasm of two adjacents cells become continuous only locally but the most of the juxtaposed cell membrane of two fusing cells still remains intact and, thus, occurrence of cell fusion could not be detected by phase contrast microscopy. Stage IV is a step in which polyerythrocyte formation can be detected by phase contrast microscopy but rounding is stopped at an intermediate stage, i.e., polyerythrocytes with a dumpbell shape. Large spherical polyerythrocytes (or polyghosts) are formed at Stage V, and this is the final step of the cell fusion reaction. (Reprinted by permission from Sekiguchi, K., Kuroda, K., Ohnishi, S., and Asano, A. (1981) *Biochim. Biophys. Acta* **645**, 211)

Fig. 3. Schematic illustration of virus envelope-cell membrane fusion.

viral contents (an RNA-protein complex called nucleocapsids) are released into the cells (Fig. 3). Concomitant with this step, macromolecules sequestered within the ghosts are released, except when a virus preparation harvested at a very early phase of infection of embryonated eggs is used. In the case of virus particles thus prepared, it has been known that freeze-thawing of such virus preparations activates macromolecule-releasing activity without affecting cell fusion-inducing activity (*13, 14*). Thus, this phenomenon may depend on the leakiness of the virus envelope. This stage is highly dependent on the reaction temperature. At 20°C and lower, almost no reaction can proceed, whereas an increase in the reaction temperature above this level up to about 40°C enhances the reaction extensively. The presence of active F glycoprotein on the viral envelope is absolutely necessary for this step.

Stage III is local fusion of the plasma membranes of adjacent cells. Although this stage has been detected by electron microscopy at an early phase of the fusion reaction (1–2 min after elevation of temperature to 28–37°C) (*15, 16*), it cannot be detected by phase contrast microscopy under usual fusion conditions. It can be measured, however, if osmotic swelling is minimized either by omitting macromolecules which are usually sequestered within the ghosts, or by adding appropriate concentrations of macromolecules to the extracellular medium. Under these circumstances, the occurrence of local cell-cell fusion can be demonstrated either by transfer of fluorescent-labeled albumin from one ghost to another, or by observation of polyghost formation after osmotic swelling in the cold with the removal of extracellular macromolecules.

Stages IV and V are rounding processes. Stage IV is a step in which polyghost formation can be detected by phase-contrast microscopy. For some reason, rounding is stopped at an intermediate stage, *i.e.*, polyghosts with a dumbbell shape, especially when a low dose of the virus is employed. Large spherical polyghosts are formed at Stage V, and this is the final step in the cell fusion reaction.

## III. STRUCTURE-FUNCTION RELATIONSHIP OF F GLYCO-PROTEIN REQUIRED FOR FUSION

The importance of F glycoprotein in the fusogenic activities of HVJ was first reported by Homma and his colleagues (*4*). During studies on host-dependent modification of HVJ, they found that virus preparations grown in L cells are different from those grown in embryonated eggs. Although these virus preparations exhibit hemagglutinating and NA activities and can infect embryonated eggs, neither fusogenic activities nor infectivity for cell cultures could be detected in such virus preparations. Comparing the chemical composition of egg-grown HVJ with L cell-grown HVJ, they found that two glycoprotein bands, $F_1$ and $F_2$,

TABLE I. N-terminal sequences of the $F_1$-polypeptides of paramyxoviruses.[a]

| | | 5 | | 10 | | 15 | | 20 | |
|---|---|---|---|---|---|---|---|---|---|
| Sendai | Phe-Phe-Gly-Ala-Val-Ile-Gly-Ile-Ile-Ala | | | | -Leu-Gly-Pro-Ala-Thr- | | | | (Ref. 22) |
| Sendai | Phe-Phe-Gly-Ala-Val-Ile-Ile-Gly-Thr-Ile-Ala-Leu-Gly-Val-Ala-Ala-Gln-Ile-Thr- | | | | | | | | (Ref. 21) |
| SV5 | Phe-Ala-Gly-Val-Val-Ile-Gly-Leu-Ala-Ala-Leu-Gly-Val-Ala-Thr-Ala-Ala-Gln-Val-Thr- | | | | | | | | (Ref. 21) |
| NDV | Phe-Ile-Gly-Ala-Ile -Ile-Gly-Gly-Val-Ala-Leu-Gly-Val-Ala-Thr-Ala-Ala-Gln-Ile-Thr- | | | | | | | | (Ref. 21) |

[a] Asterisks indicate that residues reported for the same position from different laboratories differ with each other. Dashed underlines indicate aminoacyl residues that differ in different paramyxovirus sequences.

were missing from the L cell-grown preparations, and instead a new glycoprotein band (designated as $F_0$) whose molecular weight almost corresponded to the sum of $F_1$ and $F_2$ was present in these preparations. They further found that limited proteolysis of L cell-grown samples with trypsin resulted in the activation of fusogenic activities in parallel with the splitting of $F_0$ to $F_1$ and $F_2$. As described previously, $F_1$ and $F_2$ are still interconnected with a disulfide bond(s) (8, 17). In embryonated eggs, processing of $F_0$ to $F_1$ and $F_2$ seems to occur due to endogeneous protease(s). The participation of F protein in fusogenic activities is further confirmed by either selective inactivation of the egg-grown HVJ with trypsin (18) (at much higher concentrations than that used for the activation of L cell-grown HVJ) as described below, or by simply omitting F glycoprotein from the reconstitution mixture of fusogenic proteoliposomes (19). The presence of similar F glycoproteins in other fusogenic paramyxoviruses has also been reported recently (20, 21).

The next problem is how F protein induces the membrane fusion reaction. Since the basic structure of biomembranes is lipid bilayers, the presence of lipid hydrolase activity in F protein was suspected, but no such evidence was obtained. In 1978, Gething et al. reported the amino acid sequence of the $NH_2$-terminal segment of $F_1$ (Table I) (22). To our surprise, none of these 15 amino acids from the $NH_2$-terminal has any ionizable group. Therefore it seems unusual unless this hydrophobic segment is buried within the membrane, but at least the terminal amino acid (Phe) has to be exposed to the surrounding aqueous medium, since this is the site of tryptic processing as described above.

Therefore, we started a project aimed at elucidating the following problems: 1) Is the $NH_2$-terminal segment of $F_1$ exposed to the surrounding aqueous medium? 2) Is this segment required for the fusogenic activity of the virus particles? 3) Does inactivated F glycoprotein have the $NH_2$-terminal segment buried in the hydrophobic interior of the protein or in the viral membrane? To these ends, we used limited proteolysis of the virus particles with several proteolytic enzymes, and surface iodination of the virus particles.

As summarized in Table II, some proteases preferentially affected F without attacking HANA glycoprotein. Among these, aminopeptidase M seems to digest only $F_1$, because the $NH_2$-terminals of HANA and $F_2$ are blocked (22, 23). Since the original $NH_2$-terminal residue of $F_1$, Phe, is mostly split off by this high molecular weight exopeptidase, exposure

TABLE II.   Protease digestion of HVJ glycoproteins.[a]

| | HANA | | $F(F_1)$ | | | |
| Protease | Activity | PAGE pattern | Activity | PAGE pattern | $NH_2$-terminal | [125]I-label in $F_{1a}$ |
|---|---|---|---|---|---|---|
| Trypsin | → | → | ↓ | $F_{1a}+F_{1b}$ | Phe($F_{1a}$) | ··· |
| Chymotrypsin | ↓ | ↓ | ↓ | → | Ala(Phe) | ↓ |
| Thermolysin | ↓ | ↓ | ↓ | → | Ile(Phe) | → |
| $V_8$ protease | ↓ | ↓ | ? | → | Phe | → |
| Aminopeptidase M | → | → | → | → | Ala(Phe) | ND |

[a] → no apparent change; ↓ lost. ND: not determined.

of this terminal to the aqueous environment is substantiated. The appearance of Ala as the new terminal suggests that at least three aminoacyl residues (Phe, Phe, Gly) from the $NH_2$-terminal are exposed to the aqueous medium. Although chymotrypsin and thermolysin digest both F and HANA, they preferentially inactivate F without affecting HANA at low protease concentrations (data not shown). Inspection of SDS-PAGE of these inactivated virus particles showed no appreciable difference from that of untreated virus. Therefore, we isolated F glycoprotein from the digested HVJ and compared it with intact F protein. About 3,500 and 2,500 dalton decreases in molecular weight as estimated by SDS-PAGE were noticed with chymotrypsin-inactivated and thermolysin-digested samples, respectively (data not shown). New $NH_2$-terminal residues were detected with these inactivated samples as shown in the table. Based on these results, it is tempting to speculate that the $NH_2$-terminal hydrophobic segment removed from the $F_1$ subunit by chymotrypsin or thermolysin digestion is required for fusogenic activity of the protein, and also that the segment is exposed to the surface of the protein, and therefore, preferentially split off with the proteases.

Results confirming the above hypothesis were obtained by a combination of surface [125]I-labeling by lactoperoxidase or chloramine T and proteolytic dissection. Intact virus particles are radioiodinated under conditions which are known to label only the outer surface of the virus particles. F and HANA glycoproteins and some lipid components are labeled as expected. Chymotrypsin digestion of the labeled virus particles resulted in extensive release of protein-bound [125]I. The Iodine-label present in F protein purified from chymotrypsin-treated samples was about 1/3 lower than that present in F protein prepared from untreated virus. Thermolysin digestion, however, did not result in such loss of the iodine label, although further treatment of thermolysin-digested virus particles with chymotrypsin released the iodine label as in case of direct chymotrypsin digestion. No tyrosine is present in the published 20 amino acid sequence (Table I) of the $NH_2$-terminal segment of $F_1$ (22), and more than 20 aminoacyl residues seem to be removed by chymotrypsin digestion, judging from the decrease in molecular weight ($\sim$3,500). Edman degradation was performed on F protein from the thermolysin-treated sample, in the hope of obtaining information about the iodinated site near the $NH_2$-terminal of thermolysin-digested $F_1$. Although 2–3 amino terminals were detected in each cycle, $^{125}$I-iodotyrosine was identified in second cycle of the Edman degradation. Therefore, the $NH_2$-terminal segment of $F_1$ composed of 20–30 aminoacyl residues seems to reside on the surface of F glycoprotein.

Trypsin digestion at concentrations 10 times more than those employed for activation of L cell-grown HVJ selectivity inactivated F protein functions (18). SDS-PAGE study of the digested virus particles showed that disappearance of $F_1$ and two new bands designated as $F_{1a}$ and $F_{1b}$ without affecting the other viral components. The sum of the molecular weight of $F_{1a}$ (larger one) and $F_{1b}$ (smaller one) is almost identical to that of $F_1$ under the SDS-PAGE conditions employed. Thus, it is most probable that a single split of the $F_1$ subunit resulted in complete inactivation of the fusogenic activity of F glycoprotein. Since the original $NH_2$-terminal (Phe) is retained in the $F_{1a}$ fragment, this fragment seems to correspond to the $NH_2$-terminal side of $F_1$, and $F_{1b}$ is thus assigned to the COOH-terminal side of the subunit.

Because the hydrophobic $NH_2$-terminal segment is still intact after inactivation by trypsin digestion, we need some explanation for this result. Therefore, the surface iodination technique described above was applied to this system. Trypsin-digested F did not lose the iodine label as expected, and the iodine label is almost equally distributed between $F_{1a}$ and $F_{1b}$. Trypsin splitting of iodinated, then chymotrypsin-digested virus particles similarly produced $F_{1a}$ and $F_{1b}$ fragments. But the iodine label in the $F_{1a}$ fragment is drastically decreased in this doubly-digested sample compared with that in the chymotrypsin-untreated sample. The iodination level of the other components are not appreciably different between the two samples.

Thus, the previous conclusion that release of iodine label by chymotrypsin digestion is due to loss of iodine label near the $NH_2$-terminal

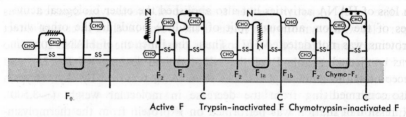

Fig. 4.   Model of several derivatives of F glycoprotein.

portion of $F_1$ is substantiated. Iodination after trypsin splitting of $F_1$ on the other hand, resulted in very low iodination of the $F_{1a}$ fragment, although the iodination level of $F_{1b}$ and the other surface components are not decreased, or in some cases, even slightly increased. This may seem peculiar at first, but can be explained easily if we assume that higher-order structures of F glycoprotein are changed after trypsin digestion, and therefore, the hydrophobic $NH_2$-terminal segment is buried in the hydrophobic interior of the protein. The above assumption was supported recently by circular dichroism measurement in the far ultraviolet region. Due to trypsin digestion, the $\alpha$-helix content of F protein is increased from 43 to 50% and that of the $\beta$-form is decreased from 12 to 9%. Contents of $\beta$-turn and unordered forms are not changed appreciably; therefore trypsin digestion seems to induce changes in the higher-order structure of F protein. Similar changes in the secondary structure were also found recently in the case of proteolytic activation of $F_0$ to $F_1$ and $F_2$ (24). Figure 4 illustrates models of the structure of several derivatives of F glycoprotein described above.

## IV.  STRUCTURE-FUNCTION RELATIONSHIP OF HANA GLYCOPROTEIN

HANA glycoprotein of HVJ was found to be disulfide-linked oligomers (presumably dimers and tetramers) (8). These disulfide bonds are unusually sensitive to low concentrations of sulfhydryl compounds, and furthermore, reductive cleavage of interchain disulfide bond(s) with reduced glutathione, dithiothreitol (DTT), or other sulfhydryl compounds is accompanied by a parallel decrease in the biological activities of HANA glycoprotein (hemagglutinating and NA activities) (8, 17). Treatment of the virus particles with these sulfhydryl compounds not only resulted in loss of HANA activities but also abolished the other biological activities of the virion, although split of disulfide bonds in the other viral proteins was not detected (17). Thus, the function of HANA protein was confirmed by these experiments to be binding to the cell surface receptor(s) (hemagglutination) and NA as previously suggested. It was also confirmed that the intactness of HANA protein is a prerequisite to expression of the other biological activities of HVJ.

Another previously undetected function of HANA protein was recently found using reconstitution of the hybrid envelope and concanavalin A (Con A)-mediated binding of HANA-inactivated virus particles to the target cells (25). For reconstitution of the hybrid envelope, we utilized the method described by Volsky and Loyter for reassembly of detergent-solubilized virus envelopes (26). Different species of virus particles or differently-treated HVJ particles were extracted with a Triton X-100-containing medium, mixed together, then dialyzed against a buffer containing SM-2 beads to remove the detergent (Triton X-100). Reassembled viral envelopes thus prepared from untreated HVJ were active for hemagglutination and hemolysis (a method for detection of virus envelope-cell membrane fusion), whereas those reassembled from trypsin-inactivated HVJ were inactive after reconstitution. Hybrid envelopes prepared from trypsin-inactivated HVJ containing inactive F and DTT-treated HVJ having inactive HANA were active for both hemagglutination and hemolysis, thus showing successful reconstitution of the hybrid envelopes.

Hybrid envelopes prepared from non-fusogenic influenza virus with HANA-inactivated DTT-treated HVJ were, however, inactive for hemolysis, although hemagglutination occurred successfully with the hemagglutinin (HA) supplied from influenza virus. Since this hybrid also contained NA originating from influenza virus, the results suggest that HA and NA of influenza virus cannot replace the function of HANA protein of HVJ in reconstitution of fusogenic envelopes. In other words, HANA protein of HVJ may participate not only in hemagglutination but also in the fusion reaction in an unknown fashion. Supporting data on this assumption were also obtained by Con A-mediated binding of DTT-inactivated HVJ to horse erythrocytes. Since horse erythrocytes have no acetylneuraminic acid (instead they have glycolylneuraminic acid) on their surface glycoproteins and glycolipids, HVJ cannot bind to these erythrocytes and therefore no reaction occurred. Addition of a limited amount of Con A, however, mediated the binding of HVJ particles to the erythrocytes, and thus hemolysis could be elicited. Split of interpeptide disulfide bond(s) on HANA protein by DTT did not affect Con A-mediated binding of HVJ to the erythrocytes, but no hemolysis could be detected with these samples. Thus, the intactness of HANA protein seems to be required for the fusion reaction. But further study will be required to elucidate the mode of participation of HANA protein in the membrane fusion reaction(s).

## V. EFFECTS OF LIPID COMPOSITION ON FUSION REACTION

Preparation of fusogenic proteoliposomes from pure viral glyco-proteins, *i.e.*, HANA and F, and chemically defined lipids was carried out recently (*19*) using a similar technique to that used for reconstitution of the hybrid envelope described above (*25*). Cell fusion reactions induced by the reconstituted system have several important characteristics similar to the virus-induced fusion reaction: fusogenic activity of the proteolipo-somes depends on the presence of active fusion protein (F) in the vesicles, and in the case of Ehrlich tumor cells, the fusion is almost completely inhibited by adding cytochalasin D to a final concentration of 4 $\mu$g/ml. Thus, the reconstituted system can be used as a model for the HVJ-induced fusion reaction.

Omission of F protein from the reconstitution mixture resulted in hemagglutinating particles with no fusogenic activity. Fusogenic activity of the reconstituted samples was highest at a HANA:F ratio of 1 : 1.5, and a protein:lipid ratio of 1 : 0.5–1.0 (w/w) gave rise to the highest fusogenic activity (M. Ozawa, unpublished observations). Although reconstitution of hemolytic vesicles was possible with a single species of phospholipid (such as phosphatidylcholine or phosphatidylethanolamine) and viral glycoproteins, no fusogenic liposomes could be obtained. On the other hand, by mixing different species of phospholipids in a ratio similar to that of the viral envelope (phosphatidylcholine:phosphatidyl-ethanolamine:phosphatidylserine:sphingomyelin = 1:2:1:1, w/w, fusogenic activity was restored when cholesterol was also included in an amount equimolar to the total phospholipids. The effect of cholesterol content on fusogenic activity was studied by varying it while fixing the interphos-pholipid and lipid-to-protein ratios. As shown in Fig. 5, cholesterol is necessary for the fusion reaction. The reason why the lipid composition

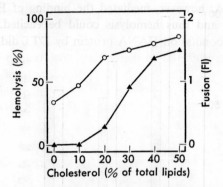

Fig. 5. Effect of choresterol content on the biological activities of the system reconstituted with a chemically defined lipid mixture containing the indicated mol percent
of cholesterol and a mixture of phosphatidylcholine, phosphatidylethanolamine,
phosphatidylserine, and sphingomyelin at a ratio of 1:2:1:1. ○ hemolytic activity;
▲ fusogenic activity expressed by fusion index (FI). (Reprinted by permission from
Ozawa, M. and Asano, A. (1981) *J. Biol. Chem.* **256**, 5954)

is so important for the fusogenic activity of proteoliposomes is unknown,
but the ability of membranes to undergo fusion may be relevant to the
microheterogeneity of lipid bilayers.

## VI.  INTRACELLULAR FACTORS WHICH CONTROL FUSION-UNDERGOING ABILITY OF CELLS

The requirement for a high intracellular ATP concentration for
HVJ-induced fusion of Ehrlich ascites tumor cells was found initially by
Okada (5) in a very early phase of his study. Conditions under which
ATP supplied only by glycolysis can support the HVJ-induced fusion
reaction have also been reported (6). The parallel decrease in fusion
frequency of Ehrlich tumor cells with intracellular ATP concentration
(manipulated by ATP-consuming reagents or ionophores) is shown in
Fig. 6. Thus, the role of ATP in regulating the fusion-undergoing ability
of cells becomes an important area to explore. Several possibilities were
tested using drugs which influence special cellular functions.

Cytochalasin B and D were found to be good inhibitors of HVJ-
induced fusion of Ehrlich tumor cells (Fig. 7) (27). The inhibition by
cytochalasin D can be assigned to its inhibitory effect on the microfilament system, since this reagent does not inhibit glucose transport of

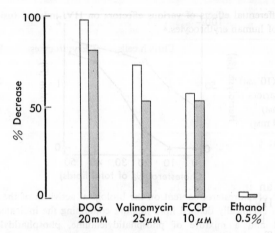

Fig. 6. ATP-depletion and inhibition of HVJ-induced fusion of Ehrlich ascites tumor cells. Open column: ATP content; dotted column: fusion frequency.

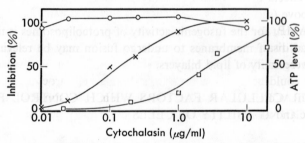

Fig. 7. Effects of cytochalasin B and D on cell fusion and intracellular ATP level. Cell fusion was performed as described previously, except that $1.9 \times 10^7$ cells and 1,000 hemagglutinating units (HAU) of the virus were used in the case of cytochalasin treatment. After cessation of the fusion reaction by cooling, a 0.5-ml aliquot of the reaction mixture of the cytochalasin D series was taken from each tube and the ATP content was determined. × cytochalasin D, fusion; ○ cytochalasin D, ATP level; □ cytochalasin B, fusion. (Reprinted by permission from Asano, A. and Okada, Y. (1977) *Life Sci.* **20**, 117)

Ehrlich tumor cells (*28*), as cytochalasin B does. The mechanism of inhibition by cytochalasins on the microfilaments system was recently revealed to be its binding to the barbed end of F-actin, and thus inhibiting end-to-side interaction of F-actin and also resulting in shorter filaments (*29*). The function of the microfilament system may be regulated by ATP. The other possibility is participation of ATP in cAMP-dependent reac-

TABLE III. Differential effects of various effectors on HVJ-induced fusion of Ehrlich tumor cells and of human erythrocytes.[a]

| Effectors | Ehrlich cells | Erythrocytes | References |
|---|---|---|---|
| ATP-depletion | ↓ | ······ | 5, 34 |
| Cytochalasin D (10 μM) | ↓ | ······ | 27, 28 |
| Mono-, di-saccharides (0.5 M) | ↓ | ······ | 33 |
| Colchicine (10 mM) | ⇑ | ⇑ | 34 |
| Theophylline (10 mM) | ⇑ | ······ | 30, 31 |

[a] ↓ inhibition; ⇑ stimulation; ······ no effect.

tions, since an increase in the intracellular cAMP level was found to enhance the HVJ-induced fusion reaction (30, 31).

As can be expected, factors which influence fusion-undergoing frequency differ depending on cell type. The biggest differences were so far observed between Ehrlich tumor cells and human erythrocytes (Table III). The effects of ATP and cytochalasin D on both systems were described above (see also Ref. 32). Theophylline treatment increased intracellular cAMP, and was stimulatory only to the Ehrlich tumor cell system. Mono- and disaccharides, such as glucose and sucrose, at high concentrations were inhibitory to Ehrlich tumor cell fusion (33), but no such inhibition could be detected in the HVJ-human erythrocyte system (data not shown). Colchicine at mM concentrations was stimulatory to both eukaryotic and akaryotic cells (34).

## VII. CHANGES IN MEMBRANE STRUCTURE RELEVANT TO CELL-CELL FUSION

With the use of freeze-fracture electron microscopy, Bächi et al. (15) observed that extensive redistribution (clustering) of intramembrane particles (IMP) occurred during HVJ-induced fusion of human erythrocytes. Such redistribution was confirmed, and this line of study was further extended by us (9, 35). Similar phenomena were also observed in the case of chemically-induced cell fusion (36, 37). Since IMP observed by freeze-fracture electron microscopy in human erythrocyte membranes have been identified as cell surface glycoproteins (38), redistribution of these glycoproteins under fusing conditions seems to be important, because they seem to be responsible for preventing close contact between plasma membranes.

Thus, we started to look for the relationship between cell-cell fusion and IMP clustering (9). Because the virus has been known to induce several changes in human erythrocytes, and only some of them may be actually related to cell fusion, critical studies will be required to draw final conclusions about the involvement of these phenomena in virus-induced fusion of erythrocyte membranes. To this end, conditions for observation of HVJ-induced fusion of human erythrocyte ghosts were developed first. Secondly, antibodies raised against purified spectrin were prepared and affinity purified. Then, antispectrin antibodies thus

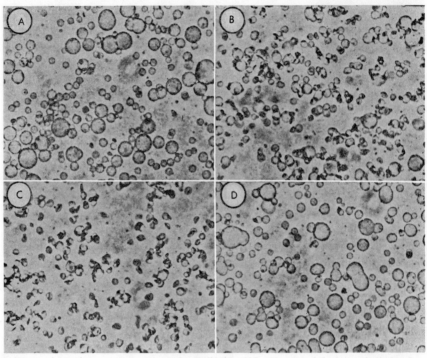

Fig. 8. Effect of antispectrin antibody on the virus-induced fusion of erythrocyte ghosts. Unsealed ghosts were resealed in isotonic buffer containing 5% bovine serum albumin and various proteins. A) Control immunoglobulin (4 mg/ml). B) Antispectrin antibody (1 mg/ml). C) Antispectrin antibody (4 mg/ml). D) $F_{ab}$ fragments prepared from the antispectrin antibody (4 mg/ml). Resealed ghosts were agglutinated by HVJ (10,000 HAU/ml) at 0°C for 15 min and then incubated at 37°C for 30 min. Phase-contrast microscopy; × 350. (Reprinted from Sekiguchi, K. and Asano, A. (1978) *Proc. Natl. Acad. Sci. U.S.* **75**, 1740)

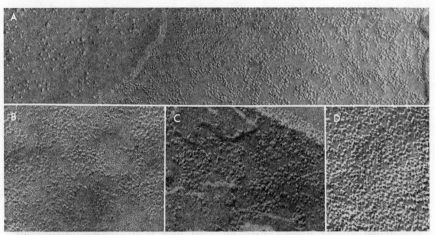

Fig. 9. Freeze-fracture picture of human erythrocyte and its ghosts, showing effect of antispectrin antibody and HVJ ($\times$ 80,000). A) PF (center to right) and EF (left) of human erythrocytes after incubation with HVJ (800 HAU/ml) for 15 min at 0°C and then for 30 min at 37°C. B), C) PF of ghosts loaded with albumin and control immunoglobulin (5 mg/ml) after incubation with HVJ (800 HAU/ml) at 0°C for 15 min and then at 37°C for 30 min. Extensive fusion in these samples was observed by phase contrast microscopy. D) Same as B and C except that the control immunoglobulin was replaced by the same concentration of the antibody. Almost no fusion was observed in this sample.

prepared were sequestered within the ghosts in the hopes of inhibiting redistribution of IMP and also inhibit HVJ-induced fusion of the ghosts as well. As shown in Fig. 8A, extensive fusion took place when bovine serum albumin(5%)-loaded ghosts, also containing 4 mg/ml of control immunoglobulin, were incubated with HVJ for 30 min at 30°C. Inclusion of antispectrin antibodies within the ghosts at 1 mg/ml (Fig. 8B) and 4 mg/ml (Fig. 8C) extensively inhibited HVJ-induced fusion. But when ghosts were loaded with $F_{ab}$ fragments of the antibodies (4 mg/ml), which can bind to spectrin only with one binding site, and thus cannot cross-link spectrin molecules, the fusion of ghosts was not affected (Fig. 8D). Therefore, it was the cross-linking of the spectrin meshwork by the divalent antibodies, rather than the binding *per se*, that was responsible for the antibody effect. Furthermore, it was recently found that anti-spectrin antibodies not only inhibited fusion of erythrocyte ghosts, but also prevented local cell-cell fusion quantitated by the transfer of fluo-rescent albumin from one ghost to another. No inhibition of viral envelope-cell membrane fusion was, however, observed with the antibody loading (K. Sekiguchi, unpublished observations).

Consequently, we examined IMP distribution during the fusion reaction of intact erythrocytes and their ghosts, and the effect of the antispectrin antibody on the distribution of the particles. The particles in the protoplasmic fracture face (PF) were dispersed in the antibody-treated ghosts, the same as in intact erythrocytes and untreated ghosts (data not shown). Extensive aggregation of IMP occurred in HVJ-induced fusion of intact erythrocytes (Fig. 9A). The particles on the external fracture face (EF) also showed non-random distribution. Furthermore, pits that might represent the place where particles were pulled out were seen on EF, and they were clustered corresponding to the distribution of the particles on the PF.

When ghosts loaded with albumin and control immunoglobulin were agglutinated with HVJ at 0°C, and then incubated at 37°C for 30 min, aggregation of IMP was observed, corresponding to extensive fusion of the ghosts (Fig. 9B and C). On the other hand, the antibodies sequestered within the ghosts almost completely inhibited the virus-induced IMP clustering under conditions that inhibited the HVJ-induced fusion reaction (Fig. 9D). Thus, the hypothesis that redistribution of IMP and concomitant formation of an area of phospholipid bilayer devoid of the particles are prerequisite for membrane fusion (15) has now obtained experimental support.

Recent studies on membrane changes in an early phase of HVJ-induced fusion of Ehrlich tumor cells by Kim and Okada (39) revealed that temperature-dependent clustering of IMP occurred on plasma membranes of fusing cells. Addition of cell fusion inhibitors, such as cytochalasin D or a high concentration of glucose, to the reaction system prevented IMP clustering as well as the cell fusion reaction. Therefore, this change in plasma membrane properties seems to be a prerequisite for the membrane fusion reaction. Since IMP clustering was temperature dependent, and since further incubation of the fusing cells decreased such clustering, this change in the Ehrlich tumor cell membrane is not permanent as in the case of erythrocyte fusion described above; rather it is more probable to assume that transient mobilization of IMP induced by HVJ is required for the fusion reaction. With the mobilization of IMP which correspond to surface glycoproteins, chances of direct contact of naked lipid bilayers may increase extensively, and thus cell-cell fusion can be initiated at these close contact areas, as in the case of erythrocyte fusion. Similar IMP clustering on the plasma membrane was also observed during polyethyleneglycol-induced cell fusion (40, 41).

Changes in the lipid bilayer structure itself which may be relevant to membrane fusion reaction were studied by physical techniques. For example, by ulitizing dilution(by cell membrane lipids)-dependent decrease of spin-spin interaction of spin-labeled phospholipids which were incorporated heavily in the virus envelope, Ohnishi and his colleagues (42) have shown that fusion of the viral envelope to the erythrocyte membrane modified the properties of the plasma membrane, and accordingly an exchange of phospholipids between the cell-surface bound but non-fusogenic viral envelope and the plasma membrane was induced. Furthermore, this modification of the membrane property seems to be propagated on the membrane (43). $^{31}$P-NMR studies on the conformational change in the membrane lipids during chemically-induced fusion of erythrocytes and liposomes seem to be a promising approach to find out what kind of modifications occurred on fusing membranes. Reports from several laboratories (44, 45) showed that a new species of lipid structure which may correspond to the hexagonal ($H_{II}$) phase (or reversed micelle) was formed during the fusion reaction induced by polyethyleneglycol, glycerol monooleate, and other chemical fusogens. The mechanism of formation of this structure is unknown at present, however.

Fusion of bare lipid bilayers is known to occur in numerous cases, provided that the lipid composition and the other conditions are favorable for the fusion reaction. In general, the presence of microheterogeneity in lipid bilayers seems to favor the occurrence of the fusion reaction between closely juxtaposed membranes. Such microheterogeneity could be induced, for example, by the presence of a high concentration of $Ca^{2+}$, the presence of some type of protein on the membrane, and so forth.

## SUMMARY

1) Reaction steps of the HVJ-induced fusion reaction were analyzed using HVJ-induced fusion of human erythrocyte ghosts as an example. Stage I is the cell agglutination step. In Stage II, fusion of the viral envelope to the plasma membrane occurs. At the next stage (Stage III),

local fusion of the adjacent plasma membrane is detected. Stages IV and V are rounding steps, and after this step detection of fusion by phase contrast microscopy becomes possible.

2)  The structure-function relationship of F glycoprotein, which is a key protein for the fusion reaction, was explored. Evidence was presented that the hydrophobic $NH_2$-terminal segment of the $F_1$ subunit of F protein participates in the fusion reaction.

3)  The structure-function relationships of HANA protein in cell agglutination and cell fusion were described. The importance of interpeptide disulfide bond(s) for biological activity and maintenance of the active structure of HANA was shown.

4)  Effects of lipid compositions including the requirement of cholesterol for the fusion reaction were described.

5)  The dependence of the fusion reaction of Ehrlich tumor cells on a high intracellular ATP concentration was discussed in relation to cytochlasin inhibition of the fusion reaction. Differences of cellular factors which control the fusion-undergoing efficiency of Ehrlich tumor cells and human erythrocytes were compared.

6)  Cluster formation of surface glycoprotein (equivalent to intramembrane partilces) was shown to be a prerequisite for cell-cell fusion of human erythrocytes. Similar mobilization of surface glycoproteins in HVJ-induced fusion of Ehrlich tumor cells and chemically-induced cell fusion was also described.

7)  The possible requirement for structural modification of lipid bilayers for the fusion reaction was discussed.

## REFERENCES

1. Poste, G. & Allison, A.C. (1973) *Biochim. Biophys. Acta* **300**, 421–465
2. Okada, Y. (1958) *Biken J.* **1**, 103–110
3. Okada, Y. (1969) *Curr. Top. Microbiol. Immunol.* **48**, 102–128
4. Homma, M. & Ohuchi, M. (1973) *J. Virol.* **12**, 1457–1465
5. Okada, Y., Murayama, F., & Yamada, K. (1966) *Virology* **28**, 115–130
6. Yanovsky, A. & Loyter, A. (1972) *J. Biol. Chem.* **247**, 4021–4028
7. Portner, A., Scroggs, R.A., Marx, P.A., & Kingsbury, D.W. (1975) *Virology* **67**, 175–187
8. Ozawa, M., Asano, A., & Okada, Y. (1976) *FEBS Lett.* **70**, 145–149
9. Sekiguchi, K. & Asano, A. (1978) *Proc. Natl. Acad. Sci. U.S.* **75**, 1740–1744
10. Sekiguchi, K., Kuroda, K., Ohnishi, S., & Asano, A. (1981) *Biochim. Biophys. Acta* **645**, 211–225

11. Markwell, M.A.K. & Paulson, J.C. (1980) *Proc. Natl. Acad. Sci. U.S.* **77**, 5693–5697
12. Markwell, M.A.K., Svennerholm, L., & Paulson, J.C. (1981) *Proc. Natl. Acad. Sci. U.S.* **78**, 5406–5410
13. Apostolov, K. & Alameida, J.D. (1972) *J. Gen. Virol.* **15**, 227–234
14. Homma, M., Shimizu, K., Shimizu, Y.K., & Ishida, N. (1976) *Virology* **71**, 41–47
15. Bächi, T., Aguet, M., & Howe, C. (1973) *J. Virol.* **11**, 1004–1012
16. Shimizu, Y.K., Shimizu, K., Ishida, N., & Homma, M. (1976) *Virology* **71**, 48–60
17. Ozawa, M., Asano, A., & Okada, Y. (1979) *J. Biochem.* **86**, 1361–1369
18. Shimizu, K. & Ishida, N. (1975) *Virology* **67**, 427–437
19. Ozawa, M. & Asano, A. (1981) *J. Biol. Chem.* **256**, 5954–5956
20. Scheid, A. & Choppin, P.W. (1974) *Virology* **57**, 475–490
21. Richardson, C.D., Scheid, A., & Choppin, P.W. (1980) *Virology* **105**, 205–222
22. Gething, M.J., White, J.M., & Waterfield, M.D. (1978) *Proc. Natl. Acad. Sci. U.S.* **75**, 2737–2740
23. Scheid, A. & Choppin, P.W. (1977) *Virology* **88**, 54–66
24. Hsu, M.-C., Scheid, A., & Choppin, P.W. (1981) *J. Biol. Chem.* **256**, 3557–3563
25. Ozawa, M., Asano, A., & Okada, Y. (1979) *Virology* **99**, 197–202
26. Volsky, D.J. & Loyter, A. (1978) *FEBS Lett.* **92**, 190–194
27. Asano, A. & Okada, Y. (1977) *Life Sci.* **20**, 117–122
28. Asano, A., Ohki, K., Sekiguchi, K., Nakama, S., Tanabe, M., & Okada, Y. (1974) *Symp. Cell Biol. (Okayama)* **26**, 39–45
29. Flanagan, M.D. & Lin, S. (1980) *J. Biol. Chem.* **255**, 835–838
30. Ohki, K., Nakama, S., Asano, A., & Okada, Y. (1975) *Biochem. Biophys. Res. Commun.* **67**, 331–337
31. Ohki, K., Asano, A., & Okada, Y. (1978) *Biochim. Biophys. Acta* **507**, 507–516
32. Miyake, Y., Kim, J., & Okada, Y. (1978) *Exp. Cell Res.* **116**, 167–178
33. Maeda, Y., Kim, J., Koseki, I., Mekada, E., Shiokawa, Y., & Okada, Y. (1977) *Exp. Cell Res.* **108**, 95–106
34. Sekiguchi, K. & Asano, A. (1976) *Life Sci.* **18**, 1383–1390
35. Asano, A. & Sekiguchi, K. (1978) *J. Supramol. Struct.* **9**, 441–452
36. Vos, J., Ahkong, Q.F., Botham, G.M., Quirk, S.J., & Lucy, J.A. (1976) *Biochem. J.* **158**, 651–653
37. Zakai, N., Kulka, R.G., & Loyter, A. (1977) *Proc. Natl. Acad. Sci. U.S.* **74**, 2417–2421
38. Shotton, D., Thompson, K., Wofsy, L., & Branton, D. (1978) *J. Cell Biol.* **76**, 512–531
39. Kim, J. & Okada, Y. (1981) *Exp. Cell Res.* **132**, 125–136
40. Knutton, S. (1979) *J. Cell Sci.* **36**, 61–72
41. Robinson, J.M., Roos, D.S., Karnovsky, M.J., & Davidson, R.L. (1979) *J. Cell Sci.* **40**, 63–75
42. Maeda, T., Asano, A., Ohki, K., Okada, Y., & Ohnishi, S. (1975) *Biochemistry* **14**, 3736–3741
43. Kuroda, K., Maeda, T., & Ohnishi, S. (1980) *Proc. Natl. Acad. Sci. U.S.* **77**, 804–807
44. Hope, M.J. & Cullis, R.R. (1981) *Biochim. Biophys. Acta* **640**, 82–90

45. Boni, L.T., Stewart, T.P., Alderfer, J.L., & Hui, S.W. (1981) *J. Membr. Biol.* **62**, 71–77

# Transmembrane Control of the Mobility of Surface Receptors by Cytoskeletal Structures

ICHIRO YAHARA

*The Tokyo Metropolitan Institute of Medical Science, Tokyo 113, Japan*

Membrane fluidity, a basic feature of biological membranes, is one of the most important concepts established during the past decade in modern cellular biology (*1*). A number of experiments have demonstrated that both lipid and protein molecules are able to diffuse laterally on or in membranes (*2*). However, recent observations on lymphocytes and other cells suggest that the cytoplasmic structures, microtubules (MT) and microfilaments (MF), regulate the mobility of surface receptors which are mostly plasma membrane glycoproteins (*3*). If the chemical and physical properties of a particular protein and lipid bilayer system and the interactions of the two components are known well enough, the distribution and mobility of the protein in the lipid bilayer is theoretically predictable. However, if the membrane is not isolated from other cellular structures and interacts not only with the aqueous medium but also with these cellular structures, these interactions must strongly affect the distribution and mobility of membrane proteins.

I deal with two subjects in this review, both of which, I believe, are essential properties of transmembrane interactions between surface receptors and cytoskeletal structures. First, negative control of the receptor mobility by cytoplasmic MT is briefly reviewed and evidence for an alteration of MT structures associated with cell surface events is presented. The second subject is positive control of the receptor mobility by the microfilamentous system. Analyses of the observations that surface microvilli appear to translocate in association with the redistribution of surface receptors are described.

## I. INTERRELATION BETWEEN CELL SURFACE RECEPTORS AND CYTOPLASMIC MT

### 1. Restriction of the Mobility of Receptors by Concanavalin A (Con A) and Its Reversion by Colchicine

In 1971, Taylor, Raff, and their associates found that binding of antibody against immunoglobulin (anti-Ig) to surface Ig on B-lymphocytes induced redistribution of surface Ig (4). After binding with divalent antibodies, the Ig-anti-Ig complexes form aggregates and subsequently results in global movement of the aggregates toward one pole of the cell (Fig. 1c). The first step of redistribution is called patch formation or patching and the second is called cap formation or capping. Patching seems to be a rather simple phenomenon in terms of molecular interactions because this is very similar to the classical antigen-antibody reaction observed in solution. Both require divalent antibody and multiple antigenic sites on the antigen. Capping appears to be more complicated, however, because this process is dependent upon cellular metabolism. In addition, it was suggested that a contractile system which consists of actin and myosin is involved in capping (4).

Edelman and I (5) found that a plant lectin, Con A, inhibited both

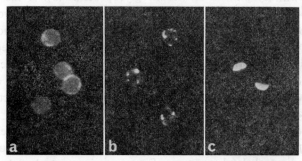

Fig. 1. Patching and capping of lymphocyte surface Ig. a) Distribution of Ig on lymphocyte surface *in situ*. b) Patchy distribution of Ig. c) Cap distribution of Ig.

patching and capping induced by anti-Ig. This effect of Con A has also been observed in the redistribution of Thy-1 antigens on mouse thymocytes, H-2 antigens on mouse lymphocytes and cultured mouse fibroblasts, and in the Con A receptors themselves. Therefore, Con A binding did not result in patching and capping. The effect of Con A is dose dependent and is effective at levels as low as 5 $\mu$g/ml.

The inhibitory effect of Con A on the redistribution of receptors is also dependent on the valence of Con A molecules (6). Native Con A is a tetramer of identical subunits under physiological conditions. But a chemically modified Con A, succinyl-Con A, is dimeric under the same conditions and possesses the same affinity for a specific sugar, $\alpha$-methyl-D-mannoside, as does Con A. However, succinyl-Con A does not inhibit patching and capping. When antibody directed against Con A was added to cells which had been previously incubated with succinyl-Con A, it inhibited patching and capping, suggesting that the cross-linking of Con A receptors is necessary for the inhibition.

The inhibitory effect of Con A was found to be suppressed when cells were preincubated with colchicine (7). Other MT-dissociating drugs, such as colcemid, vinblastine, vincristine, and podophyllotoxin, show similar effects. An isomer of colchicine, lumicolchicine, which has been shown not to dissociate MT, shows no effect. In addition, the reversion of the Con A effect was also observed when cells were preincubated with Con A at 4°C where MT are known to be depolymerized (8). Con A itself can induce capping under conditions where the inhibitory Con A effect on the redistribution of receptors is suppressed. Electron microscopic observations on ultrathin sections and ghost-membranes of cells incubated with ferritin-labeled Con A indicate that patching is induced on colchicine-treated cells but not on untreated cells (9). From the above observations, it was concluded that MT or related structures are involved in the restriction of mobility of receptors by Con A. Particularly, it must be noted that the regulation of receptor mobility by cooperative action of Con A and MT occurs at the individual receptor level.

MT structures 25 nm in diameter are not always situated just beneath the plasma membranes, however (9). In contrast, MF 5–8 nm in diameter have always been observed under the plasma membrane of mouse lymphocytes (9, 10). These results suggest that the interaction between surface receptors and MT may be mediated by cortical MF.

Using Con A-coated nylon fibers (11), Con A-bound platelets (12), and Con A-coated latex beads (12), it was demonstrated that Con A locally bound to cells is enough to inhibit the mobility of receptors on the whole cell surfaces of these cells. When lymphocytes were preincubated or post-treated with colchicine, the inhibition of patch and cap formation was suppressed and, in addition, bound Con A-coated particles accumulated at one pole of the cell (12). The effect of locally bound Con A on the receptor mobility was verified later with fibroblasts by means of fluorescence photobleaching recovery, a physico-chemical method for determining receptor movements (13).

## 2. A Hypothesis for the Control of Receptor Movement by Cytoplasmic MT

Using the observations described above, Edelman and I (3, 9, 12) have proposed the working hypothesis that cell surface receptors are anchored on a common assembly which is in or under the plasma membrane. The detailed hypothesis incorporates the following assumptions.

a)   Certain surface receptor molecules penetrate the lipid bilayer of the plasma membrane and interact with other cellular structures in or under the plasma membrane. Such surface receptors have been found in erythrocytes (14) and lymphocytes (15).

b)   Some of these surface receptors interact reversibly and probably indirectly with the MT organization of the cytoplasm. We call A the anchored state of the receptors which are attached to MT and F the state of the receptors which are free from the MT; these two states are assumed to exist in an equilibrium A⇌F. Through this anchorage, the distribution of the receptors on the cell receptor is affected by the state of the MT.

c)   Conversely, the state of MT organization is affected by cross-linking interaction and aggregations of particular receptors. This provides a means by which receptor states can be communicated to the interior of the cell. The valence of external ligands can therefore be a critical factor in cell surface-cytoplasmic interactions.

d)   The mobility of the membrane or its receptors is affected by the state of MT, and therefore alteration of MT by one set of cell surface receptors may affect the movement of the other receptors.

e)   Finally, the equilibrium between the two states (A and F) of the receptors is affected by colchicine and related agents. Alteration of the equilibrium may occur either because structures such as MT are dissociated by these agents or because receptors are released from attachment with the MT assembly, or both.

### 3. Visualization of the MT Network in Lymphocytes by Immunofluorescence Microscopy Using Antitubulin Antibody

As to the hypothetical model described above, two key assumptions should be tested experimentally. First, an assumption that the state of surface receptors alters in association with the state of cytoplasmic MT has already been proven by the observation that surface receptors altered their distribution and mobility when cells were treated with colchicine. The second is a problem as to whether cell surface events concerning receptors induce an alteration of MT organization. An attempt has been made to compare the MT organization or normal and modulated cells on which cell surface events such as capping were induced (9). However, we have encountered difficulties in observing MT in most populations of lymphocytes by electron microscopy using thin sections, although cytoplasmic MT could be detected occasionally. An improvement of the immunofluorescence method (16) using antitubulin antibody enabled us to test this problem critically.

Monospecific antitubulin antibody eluted from a Sepharose-tubulin

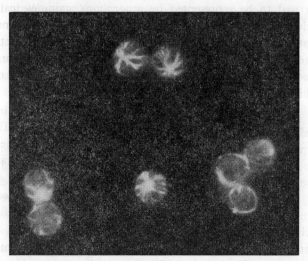

Fig. 2. MT organization of mouse lymphocytes visualized by the indirect immunofluorescence method using antitubulin antibody.

column was used for immunofluorescence staining (*17*). Formaldehyde-fixed lymphocytes were firmly attached to microscopic glass slides by cytocentrifugation. The cells were labeled with antitubulin antibody as previously described (*18*). A typical view of the sample under a fluorescence microscope is shown in Fig. 2. In most of the cells tubular structures revealed by fluorescence were detected, and each tubule appeared to originate from a structure that had been considered to be an organization center or a related structure (*17*).

Inasmuch as the antibody preparation used in our study has been proven to be specific to tubulin, we concluded that the stained structures of lymphocytes represented in Fig. 2 could be MT and related structures. The number of tubular structures revealed by immunofluorescence was estimated to be 5–20 per lymphocyte. It is not evident that a tubular structure corresponds to an individual MT. Some of these tubular structures could be bundles of a few or several MT. Microtubules appeared to surround the nucleus. It might be noted that amorphous stainings with antitubulin were observed only very weakly in the cytoplasm of stained cells. Since at least 90% of small and medium lymphocytes revealed fine MT organization, both T- and B-lymphocytes could have well-organized MT. When lymphocytes were treated with colchicine before fixation with formaldehyde, the MT organization was found to disappear and to be represented by amorphous stainings instead. An organization center-like structure was visualized much more clearly after treatment with colchicine, probably because the structure was resistant to colchicine. Microtubules were disorganized when lymphocytes were exposed to a temperature of 4°C, but regained their organized structures within 5 min after a shift of temperature to 30°C.

## 4.  Modulation of MT Organization by Patching and Capping

Lymphocytes were induced to form caps by incubation with rhodamine-labeled goat antibody against mouse IgG (rh-G-anti-MIg). The cells were then treated with antitubulin and fluorescein-labeled goat antibody against rabbit IgG (fl-G-anti-RIg) to visualize the tubulin distribution. Cells showing cap distribution of surface Ig were arbitrarily chosen with the aid of rhodamine labeling (Fig. 3a). It was found that MT organizations seen in untreated lymphocytes were absent in cells showing capping with rh-G-anti-MIg. In addition, each cap region was observed to be more brightly stained with tubulin antibody than other

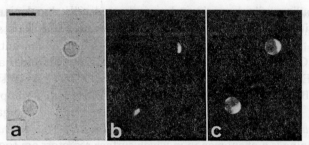

Fig. 3. Alteration of MT organization by capping. Lymphocytes incubated with rh-G-anti-MIg were examined for distribution of tubulin by the indirect immunofluorescence method using antitubulin antibody and fl-G-anti-RIg. a) Phase contrast image. b) Ig distribution visualized by rhodamine labeling. c) Tubulin distribution visualized by fluorescence labeling. Bar=10 μm. (From Ref. *17*)

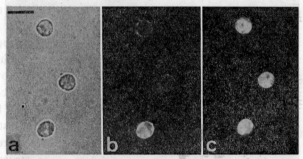

Fig. 4. Disordered MT organization of lymphocytes showing patching. Lymphocytes were incubated with rh-G-anti-MIg in the presence of sodium azide and then labeled with antitubulin antibody and fl-G-anti-RIg. a) Phase contrast image. b) Ig distribution visualized by rhodamine labeling. c) Tubulin distribution visualized by fluorescein labeling.

regions of the same cells. Staining patterns of cap regions with tubulin antibody might not be amorphous, but could not be analyzed in detail due to the limited resulotion of fluorescence microscopy.

Almost all of the B-lymphocytes began to show a patchy distribution of Ig when incubated with rh-G-anti-MIg in a medium containing sodium azide, a metabolic inhibitor, which is known to allow patching but to prevent capping (*4*). The cells showing patchy distribution of Ig were then examined for the distribution of tubulin. As can be seen in Fig. 4, B-cells showing Ig patches exhibited relatively disordered staining patterns with tubulin antibody compared to untreated cells. Incubation of cells with sodium azide alone did not cause any alteration of MT organization.

## II. MECHANISM OF CAPPING: AN ALTERATION OF MICRO-FILAMENTOUS ORGANIZATIONS ASSOCIATED WITH RE-DISTRIBUTION OF SURFACE RECEPTORS

### 1. Ligand-independent (LI) Capping Induced in Hypertonic Medium

Capping is a phenomenon in which surface receptors are coordinately collected at one pole of the cell as a result of signals such as the binding of multivalent ligands (Fig. 1). Two lines of evidence suggest that capping

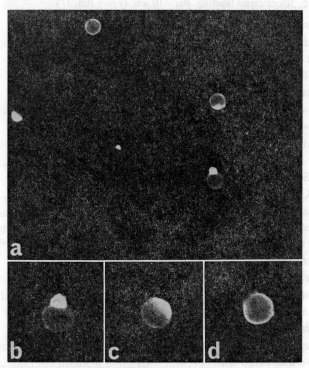

Fig. 5. Distribution of surface Ig on lymphocytes incubated in a hypertonic medium without addition of anti-Ig. a) Three caps, one intermediate type, and one diffuse distribution. b) Cap distribution. c) Intermediate distribution. d) Diffuse distribution. Lymphocytes were incubated at 37°C for 15 min in 2×PBS, fixed, and labeled with fl-R-anti-MIg.

involves a contractile system consisting of actin, myosin, and regulating proteins: a) capping is inhibited by cytochalasins which are known to interfere with the function of MF and to disorganize MF structures (4, 10); and b) actin cocaps with cell surface receptors (19, 20).

Recently, Kakimoto-Sameshima and I have found that most of the mobile receptors on mouse splenic lymphocytes and thymocytes were induced to form caps when these cells were incubated in a hypertonic medium without the addition of a ligand (21, 22). This system, designated as LI-capping, can be used to analyze structural changes in lymphocytes undergoing capping because a large number of receptors on each cell are nonspecifically induced to form caps and, therefore, changes in cellular structure associated with capping are expected to be easily detected. Mouse lymphocytes were incubated at 37°C for 10 min in 2 × PBS, which is composed of 2-fold amounts of each component of phosphate-buffered saline and has an osmolarity of 600 mOsm, and fixed with formaldehyde. The fixed cells were labeled with fl-anti-MIg. Up to 70% (but generally around 50%) of the labeled cells showed caps (Fig. 5). Incubation of lymphocytes in PBS containing 0.2–0.4 M sucrose without the addition of a ligand also resulted in capping of surface Ig. About 500–600 mOsm medium was found to induce optimal LI-cap formation.

In addition to surface Ig on B-cells, H-2 antigens on lymphocytes and thymocytes and Thy-1.2 antigens on thymocytes were found to cap in 2 × PBS without the addition of any antibody. Parts of the populations of both Con A receptors and antigenic sites for antimouse thymocyte sera on each cell also appeared to cap in 2 × PBS. Although a variety of surface receptors were induced to form caps by incubation in hypertonic medium, the inducible mobilities of different receptors were found to differ. Surface Ig molecules were more easily induced to form LI-caps than Con A receptors, suggesting that the former receptors were more mobile than the latter when cells were incubated in 2 × PBS (22). This seems to suggest that Ig might interact more strongly with the hypothetical machinery of capping in 2 × PBS than Con A receptors. It may be noted that intramembrane particles visualized by the freeze fracture method were not redistributed in cells showing LI-capping (23).

One of the feature of LI-capping is the association of caps with surface microvilli (21). During LI-capping, surface microvilli appeared to translocate together with surface receptors. Analyses of the translocation of microvilli will be described later.

TABLE I.   Inhibition of LI-capping in $2 \times$ PBS.

| Inhibitor | Inhibition (%) |
|---|---|
| 20 mM NaN$_3$ | 98 |
| 10 $\mu$g/ml oligomycin | 99 |
| 10 $\mu$g/ml CB | 100 |
| $10^{-4}$ M colchicine | 0 |
| 100 $\mu$g/ml Con A | 95 |
| 5 mM EDTA | 0 |
| 1 mM EGTA | 0 |
| 2 mM cAMP | 0 |
| 2 mM db-cAMP | 0 |
| 2 mM cGMP | 0 |
| $10^{-5}$ M carbamylcholine | 0 |
| $2 \times 10^{-4}$ M chlorpromazine | 100 |

To compare LI-capping with ligand-dependent (LD) capping, the requirements for LI-capping in hypertonic medium were tested (24). LI-capping was strongly inhibited by 20 mM sodium azide, 10 $\mu$g/ml of cytochalasin B (CB), or 100 $\mu$g/ml of Con A (Table I). Colchicine (0.1 mM) partially reversed the inhibitory effect of Con A on LI-capping, suggesting that colchicine-sensitive structures are involved in the inhibition of receptor mobility by Con A in the LI-capping system as well as in anti-Ig-induced patch and cap formation (21). Sensitivities of LI-capping to the drugs tested were found to be generally similar to those of LD-capping with the exception of the sensitivity to CB. While CB has been reported to inhibit LD-capping partially (4, 10), the drug was observed to inhibit LI-capping completely at a concentration of 10 $\mu$g/ml.

Bipolar cap formation was found in a minor population of lymphocytes which were incubated in $2 \times$ PBS in the absence of a ligand. Bipolar capping was observed both with surface Ig and with Con A receptors. The relative positions of the two caps on each of these cells appeared to be determined randomly, for the two caps were adjacent to each other on some cells and, on the other hand, were located opposite to each other on other cells (22). Although bipolar capping was detected only in a minor population of the total cells, no cell showing a bipolar cap has been found at all in the case of standard capping that generally requires specific ligands. Furthermore, it was shown in a few instances that the Ig-cap was unipolar whereas the cap of Con A receptors was bipolar on the same cells when incubated in $2 \times$ PBS in the presence of anti-Ig. These results strongly suggest that the formation of bipolar caps is not only due to incubation of the cells in a hypertonic medium but is also due to the absence of ligand in $2 \times$ PBS. Bipolar caps observed in hypertonic medium are not a transient state during the formation of standard unipolar caps, because kinetic studies of bipolar capping indicate that cells showing bipolar caps did not decrease even after incubation at 37°C for 1 h, whereas LI-capping reached a plateau within 15 min.

Caps formed in a hypertonic medium in the absence of ligand revert to the original diffuse distribution when the cells are transferred to an isotonic medium. Neither a metabolic inhibitor, sodium azide, nor CB inhibits the reversal of capping, suggesting that the reversal process might be directed by the diffusion of receptors on the cell surface. The diffusion rate of free surface proteins (see discussions in Ref. 25) can account for the kinetics of the reversal of capping. It has also been shown that the maintenance of caps required functions which are sensitive to metabolic inhibitors or to CB. In addition, a certain population of lymphocytes is able to form caps and revert to showing diffuse distribution of receptors repeatedly by altering the osmolarity of the medium from isotonic to hypertonic and *vice versa*. These results suggest that the hypothetical machinery of capping might be activated in a hypertonic medium and its activation might cease in an isotonic medium or by blocking either the cellular metabolisms or functions of MF. It is possible that LI-capping of lymphocytes takes place as a result of reversible interactions between surface receptors and an activated contractile system of these cells. Contractile forces which are required for capping are probably provided by interactions of actin filaments (MF) with myosin, as is the case for the contraction of striated muscles. In fact, it has been recently demonstrated that cross-linked surface Ig was linked to actin in intact cells (26).

## 2. Mechanism of Translocation of Microvilli Accompanying Capping of Surface Receptors

A prominent feature of LI-capping in a hypertonic medium is the accumulation of microvilli in the cap region (Figs. 5 and 7j, k). The cellular machinery directing the translocation of microvilli along the cell surface seems to be related to that of surface receptor capping.

There are three possible models to explain the translocation of microvilli along the cell surface. Model A: Microvilli move laterally without change in shape. Model B: Microvilli retract and then are reformed in other places on the cell surface. Model C: microvilli are transformed into other structures which move on the cell surface and the reverse transformation into microvilli occurs after translocation. To test these models, time-dependent alterations in the surface morphology after cells were transferred to 2 × PBS were determined by scanning electron microscopy (24).

The time course of LI-capping of surface Ig after the shift in the

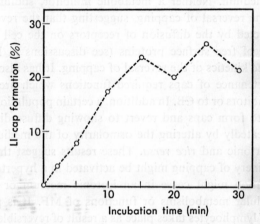

Fig. 6.  Time course of LI-capping. Lymphocytes were incubated in 2 × PBS at 37°C for the indicated periods.

TABLE II.  Morphological analysis of lymphocytes undergoing LI-capping. Lymphocytes were incubated in 2 × PBS at 37°C for the indicated periods and then processed for observation by SEM. A hundred cells were examined for each incubation.

| Group | Designation | Incubation time in 2 × PBS (min) | | | | |
|---|---|---|---|---|---|---|
| | | 0 | 5 | 10 | 15 | 30 |
| | | Number of lymphocytes | | | | |
| 1 | Smooth surface | 6 | 3 | 2 | 7 | 8 |
| 2 | Microvilli | 91 | 10 | 31 | 15 | 32 |
| 3 | Lamellae | 3 | 81 | 49 | 47 | 29 |
| 4 | Accumulated micro-villi and lamellae | 0 | 6 | 17 | 22 | 15 |
| 5 | Accumulated microvilli and smooth surface | 0 | 0 | 1 | 9 | 16 |

medium osmolarity was determined. LI-capping reached a plateau within
15 min (Fig. 6). To determine the correlation between the movement of
surface receptors and the translocation of microvilli, we incubated
lymphocytes in $2 \times$ PBS for various periods and then made a morpho-

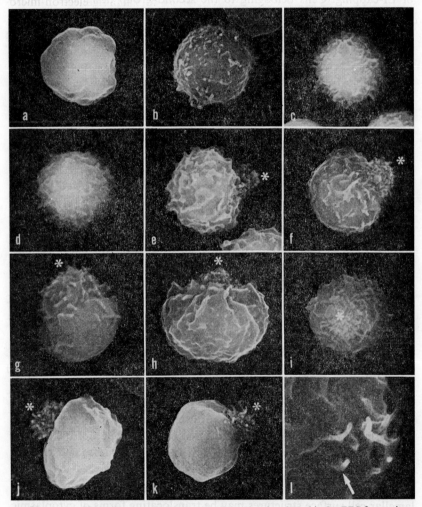

Fig. 7.    Scanning electron micrographs of lymphocytes incubated in $2 \times$ PBS for various
periods. Asterisks indicate the location of accumulated microvilli. Magnification:
a–f, h–j, $\times$ 3,770; g, $\times$ 4,700; k, 3,200; l, $\times$ 13,800.

logical analysis by scanning electron microscopy (SEM) which we compared with the time course of capping shown in Fig. 6 (24). The surface
morphologies of these cells were classified into five groups for each
incubation (Table II). A minor population of cells with a smooth surface
(Fig. 7a) does not appear to be related to the translocation of microvilli
of LI-capping. Cells belonging to the second group designated "microvilli" (Fig. 7b) were predominant among cells which were incubated in
PBS. When incubated in $2 \times$ PBS at 37°C for 2 to 5 min, cells covered
with lamellae on their surface (group 3) were predominantly observed
(Fig. 7d). Lamellar structures observed by SEM were obviously distinguishable from the microvillous structures and varied in shape and
size. Extensively well-developed lamellae were observed to be 1 $\mu$m wide,
0.5 $\mu$m high, and 0.1 $\mu$m thick. This suggests that the disappearance of
microvilli and the appearance of lamellae were almost simultaneous on
the same lymphocytes shortly after the medium shift. Both microvilli
and lamellae were detected together on some lymphocytes (Fig. 7c).
These cells were classified into group 2 or 3 according to the predomin-
ance of either structure. In addition to the coexistence of microvilli and
lamellae on the surface of the same cells, protrusions that could not be
identified easily as microvilli or lamellae were frequently observed on the
surface of cells that had been incubated in $2 \times$ PBS for short periods.
Some of these structures appeared to be microvillous at their distal end
but lamellar at their proximal part as indicated by an arrow in Fig. 7l.

Cells with accumulated microvilli and a dispersed distribution of
lamellae (group 4: Fig. 7e–i) appeared 5 min after incubation in $2 \times$ PBS
as surface receptors moved to form a cap at the same pole. The numbers
of lamellae on these cells were fewer than those on cells without accumulated microvilli (Fig 7d). Cells with accumulated microvilli and smooth
surfaces (group 5: Fig. 7j, k) appeared after incubation for 15 min or
longer. Observation of a considerable population of cells belonging to
group 4 showed that lamellae are not evenly distributed over the entire
cell surface. Most lamellae are distributed near the region of the cap but
the area opposite the cap was devoid of lamella, as if these lamellae were
moving toward the cap region (Fig. 7f–h).

These results seem to follow Model C, and they suggest that the
lamellar cell surface structures may be translocating forms of cytoplasmic
components that construct microvilli in their resting state. The disappearance of microvilli and the appearance of lamellae occurred almost

simultaneously on the same lymphocytes shortly after the shift of the medium, as if the occurrence of the latter phenomenon depends upon that of the former. Both surface structures, microvilli and lamellae, are composed mainly of MF, therefore, they would be convertible from one structure to the other.

This is compatible with the observations made by others that a) lamellar structures are rich in B-lymphocytes that react with anti-Ig whereas short microvilli are rich in untreated B-cells (27); b) retraction of microvilli is followed by formation of lamellar structures during attachment and spreading of trypsinized mouse fibroblasts on the substratum (28); and c) microvilli are developed to a greater extent on metabolically inactive cells than on active cells (29, 30). Thus, according to Model C, microvilli are structures that are not activated. If microvilli are converted into lamellae and *vice versa* on the cell surface, this corresponds to the conversion between MF bundles and their networks in the cytoplasm.

Cocapping of surface receptors and cytoplasmic contractile proteins such as actin and myosin has been shown on lymphocytes and other cells (19, 20, 31). A similar cocapping occurs in LI-capping (24).

If Model C does represent the actual mechanism of microvilli translocation, then the simultaneous movements of surface receptors and microvilli might be explained by the hypothetical interaction of receptors with lamellae. Although only cross-linked receptors are able to interact with lamellae in LD-capping in an isotonic medium, most mobile receptors can interact in LI-capping in a hypertonic medium. Recently, histocompatibility antigens and cross-linked surface Ig on lymphocytes and other cells have been shown to be associated with actin (26), which is consistent with the hypothetical interaction described above. Furthermore, indirect evidence suggests that the movement of particles attached to the dorsal surface of the leading edges of fibroblasts is attributable to the motion of lamellar structures (32). Taken together, all this evidence shows that it is likely that surface lamellae, not microvilli, are responsible for the translocation of receptors during capping.

## SUMMARY

Transmembrane interaction between surface receptors and cytoskeletal structures is a prominent feature of the plasma membrane that surrounds the whole cell body. In contrast to proteins included in isolated

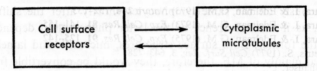

Fig. 8.   Interaction between cell surface receptors and cytoplasmic MT. If one of the two is changed, the other is affected.

or reconstituted membranes that are essentially diffusible along these membranes, plasma membrane proteins of intact cells are restricted from moving in some cases, and are also facilitated to translocate in other cases by the aid of cellular machinery which requires metabolic energy and a contractile system.

Two subjects are dealt with in this review. First, previous studies have suggested that MT are involved in the restriction of receptor mobility. In this system, not only are the mobility and distribution of surface receptor proteins affected by the state of MT organization, but conversely, the state of MT organization is also affected by cell surface events such as cross-linking of receptors (Fig. 8). Evidence for an interrelation between surface receptors and the cytoplasmic MT organization is presented and discussed.

Second, although a contractile system, microfilamentous structures, has been suggested to be involved in surface receptor capping, the mechanism of capping remains to be clarified. The finding of an intriguing phenomenon, ligand-independent-capping in a hypertonic medium, provided an opportunity to investigate the relation between alterations in surface architectures and translocation of receptors. The conversion of microvilli consisting of MF bundles into lamellae consisting of MF networks is suggested to be involved in the translocation of receptors along membranes.

## REFERENCES

1. Singer, S.J. & Nicolson, G.L. (1972) *Science* **175**, 720–731
2. Edidin, M. (1974) *Annu. Rev. Biophys. Bioeng.* **3**, 179–201
3. Yahara, I. & Edelman, G.M. (1975) *Ann. N.Y. Acad. Sci.* **253**, 455–469
4. Taylor, R.B., Duffus, W.P.H., Raff, M.C., & de Petris, S. (1971) *Nature New Biol.* **233**, 225–229
5. Yahara, I. & Edelman, G.M. (1972) *Proc. Natl. Acad. Sci. U.S.* **69**, 608–612
6. Gunther, G.R., Wang, J.L., Yahara, I., Cunningham, B.A., & Edelman, G.M. (1973) *Proc. Natl. Acad. Sci. U.S.* **70**, 1012–1016

7. Yahara, I. & Edelman, G.M. (1973) *Nature* **246**, 152–155
8. Yahara, I. & Edelman, G.M. (1973) *Exp. Cell Res.* **81**, 141–155
9. Yahara, I. & Edelman, G.M. (1975) *Exp. Cell Res.* **91**, 125–142
10. de Petris, S. (1977) *Cell Surf. Rev.* **3**, 643–728
11. Rutishauser, U., Yahara, I., & Edelman, G.M. (1974) *Proc. Natl. Acad. Sci. U.S.* **71**, 1149–1153
12. Yahara, I. & Edelman, G.M. (1975) *Proc. Natl. Acad. Sci. U.S.* **72**, 1579–1583
13. Schlessinger, J., Elson, E.L., Webb, W.W., Yahara, I., Rutishauser, U., & Edelman, G.M. (1977) *Proc. Natl. Acad. Sci. U.S.* **74**, 1110–1114
14. Bretscher, M.S. (1971) *Nature New Biol.* **231**, 229–232
15. Walsh, F.S. & Crumpton, M.J. (1977) *Nature* **269**, 307–311
16. Lazarides, E. & Weber, K. (1974) *Proc. Natl. Acad. Sci. U.S.* **71**, 2268–2272
17. Yahara, I. & Kakimoto-Sameshima, F. (1978) *Cell* **15** 251–259
18. Weber, K., Bibling, T., & Osborn, M. (1975) *Exp. Cell Res.* **95**, 111–120
19. Gabbiani, G., Chaponnier, C., Zumbe, A., & Vassalli, P. (1977) *Nature* **269**, 687–688
20. Sundqvist, K.G. & Ehrnst, A. (1976) *Nature* **264**, 226–231
21. Yahara, I. & Kakimoto-Sameshima, F. (1977) *Proc. Natl. Acad. Sci. U.S.* **74**, 4511–4515
22. Yahara, I. & Kakimoto-Sameshima, F. (1979) *Exp. Cell Res.* **119**, 237–252
23. Yahara, I. & Kakimoto-Sameshima, F. (1980) *Cell Struct. Funct.* **5**, 223–232
24. Yahara, I. & Kakimoto-Sameshima, F. (1979) *Cell Struct. Funct.* **4**, 143–152
25. Bretscher, M.S. (1976) *Nature* **260**, 21–23
26. Flanagan, J. & Koch, G.L.E. (1978) *Nature* **273**, 278–281
27. Kay, M.M.B. (1975) *Nature* **254**, 424–426
28. Arbrecht-Buehler, G. (1976) *Cell Motility* (Goldman, R., Pollard, T., & Rausenbaum, J., eds.) pp. 247–264, Cold Spring Harbor Laboratories, New York
29. Loor, F. & Hagg, L.-B. (1975) *Eur. J. Immunol.* **5**, 854–865
30. de Petris, S. (1978) *Nature* **272**, 66–68
31. Schreiner, G.F., Fujiwara, K., Pollard, T.D., & Unanue, E.R. (1977) *J. Exp. Med.* **145**, 1393–1398
32. Abercrombie, M., Heaysman, J.E.M., & Pregrum, S.M. (1970) *Exp. Cell Res.* **62**, 389–398

# Structures of the Sugar Chains of Cell Surface Glycoproteins

AKIRA KOBATA

*Department of Biochemistry, Kobe University School of Medicine, Kobe 650, Japan*

In recent years, studies of the structures and functions of complex carbohydrates have attracted the interest of many biochemists. This is because various studies in the field of cell biology have suggested that complex carbohydrates play important roles in cell surface recognition phenomena. Early in 1964, Gesner and Ginsburg (*1*) studied the lymphocyte homing mechanism, that is, the ability of lymphocytes to pass through the endothelial cells lining blood vessels, and found that exoglycosidase treatment almost completely deprives the cells of their homing activity. Since then, various kinds of biological processes such as specific cell-to-cell adhesion, differentiation of animal cells, and specific mating reactions of gamete cells have been shown to include cell surface glycoproteins. In order to understand the molecular basis of these interesting biological phenomena, information on the structures of complex carbohydrates, especially on their sugar chain moieties, are indispensable. Because glycoproteins usually contain several sugar chains with different structures, most structural studies of the carbohydrate moiety have been performed by using glycopeptides obtained by exhaustive pronase digestion of glycoproteins (*2*). This routine method, however, often led to false conclusions about the sugar chain structures, because of the structural diversity of the peptide moieties. The establishment of enzymatic and chemical methods to release the sugar chains almost intact from the polypeptide backbone has opened a new age in the structural study of glycoproteins. The sugar chains of glycoproteins can be classified into two groups according to the structures of the regions linking them to the polypeptide backbone. In one group, the sugar chains are attached to the polypeptide by an O-glycosidic linkage from N-acetylgalactosamine to serine or threonine. Since this group is found abundantly in mucins, such sugar chains are arbitrarily called mucin-type sugar chains. Another group of sugar chains are linked N-glycosidically from N-acetylglucos-amine to the amide nitrogen of asparagine in the peptide and are called asparagine-linked sugar chains.

## I.  STRUCTURES OF MUCIN-TYPE SUGAR CHAINS

Since the sugar chains of this group are easily released from the polypeptide backbone by the $\beta$-elimination reaction which is performed by mild heating in alkaline borohydride solution (3), their structures have been most conveniently studied by using oligosaccharides obtained through the reaction. The linkage N-acetylgalactosamine is converted to N-acetyl-galactosaminitol and can be identified after acidic hydrolysis of the released oligosaccharides. If $NaB[^3H]_4$ is added to the $\beta$-elimination reaction mixture, oligosaccharides with N-acetyl[1-$^3$H]galactosaminitol at their reducing termini are obtained, making their detection easier and more sensitive. The structures of some of the mucin-type sugar chains thus determined are listed in Table I. Structures I, II, and IV have been found most widely in nonmembrane and membrane glycoproteins. Structure I is a kind of common core possessed by most mucin-type sugar chains, and it is also derived from porcine blood group H and A substances by the addition of fucose, N-glycolylneuraminic acid, and N-acetylgalactosamine (13). In some cases, this core is elongated by the addition of Gal$\beta$1→GlcNAc$\beta$1→3 repeating units (Tables I, VI). An extremely large mucin-type sugar chain was shown to occur in the blood group substance obtained from human ovarian cysts (Tables I, VII).

## II.  STRUCTURES OF ASPARAGINE-LINKED SUGAR CHAINS

Because of the lack of a convenient method to release the sugar

**TABLE I.**  Sugar chains linked through $N$-acetylgalactosamine to the hydroxyl group of serine and threonine residues.

| | Structure | Source of glycoproteins |
|---|---|---|
| I. | Galβ1→3GalNAc→Ser(Thr) | hCG[a] (4), RBC[b] (5), human IgA (6), antifreeze glycoprotein of antarctic fish (7) |
| II. | Galβ1→3GalNAc→Ser(Thr)<br>　　　　3<br>　　　　↑<br>　　NeuAcα2 | RBC (5), bovine kininogen (8) |
| III. | GalNAc→Ser(Thr)<br>　　　6<br>　　　↑<br>NeuAcα2 (NeuGly) | Submaxillary mucins (9) |
| IV. | Galβ1→3GalNAc→Ser(Thr)<br>　　　　3　　　　　6<br>　　　　↑　　　　　↑<br>　NeuAcα2　　NeuAcα2 | RBC (5), bovine kininogen (8) |
| V. | Galβ1→3GalNAc→Ser(Thr)<br>　　　2<br>　　　↑<br>　Fucα1 | RBC (10) |
| VI. | Galβ1→4GlcNAcβ1<br>　　　　　　　　　↘6<br>Galβ1→3GlcNAcβ1→3Galβ1→3GalNAc→Ser(Thr)<br>　　　　　　　　　　　　　　　　　6<br>　　　　　　　　　　　　　　　　　↑<br>　　　　　　　　　　　　　Galβ1→4GlcNAcβ1 | Human gastric mucin (11) |
| VII. | 1βGlcNAc3←1βGal<br>　　　　　　4<br>　　　　　　↑ 6<br>Galα1→3Galβ1→4GlcNAcβ1↘<br>　3<br>Galα1→3Galβ1→3GlcNAcβ1↗<br>Galβ1→4GlcNAcβ1<br>Fucα1　Fucα1<br>　2　　　3<br>Galα1→3Galβ1→4GlcNAcβ1<br>　2　　　3<br>Fucα1　Fucα1<br>Galα1→3Galβ1→3GlcNAcβ1<br>　2　　　4<br>Fucα1　Fucα1 | Human ovarian cyst (B-type) (12) |

[a] hCG: human chorionic gonadotropin.  [b] RBC: human red blood cells.

chain moiety from the polypeptide backbone, elucidation of the structures of asparagine-linked sugar chains was slow as compared to that of
mucin-type sugar chains.

However, the discovery of endo-β-N-acetylglucosaminidases (14),
which cleave the asparagine-linked sugar chains mostly intact from the
polypeptide backbone, and the establishment of hydrazinolysis, a chemical method which specifically cleaves the GlcNAc→Asn linkage (15),
have promoted development in this field of research. Structurally, these
sugar chains can be classified into three subgroups (Fig. 1).

Those in the high mannose-type group contain only mannose and
N-acetylglucosamine residues. They contain a heptasaccharide, Manα1→
6(Manα1→3)Manα1→6(Manα1→3)Manβ1→4GlcNAcβ1→4GlcNAc, as
a common core, and variation is formed by the number of Manα1→2
residues linked to the three nonreducing terminal α-mannosyl residues
of the core portion.

Those which fall into the second subgroup, the *complex type*, contain
a pentasaccharide, Manα1→6(Manα1→3)Manβ1→4GlcNAcβ1→4Glc
NAc, as a common core and variation occurs in the number of outer chain
moieties linked to the two α-mannosyl residues of the core and in many
cases an α-fucosyl residue is linked at the C-6 position of the proximal

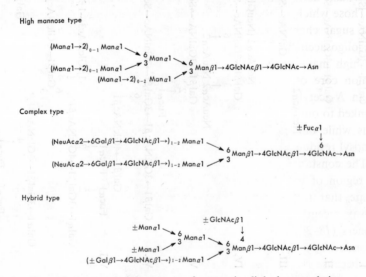

Fig. 1.   General structures of three types of asparagine-linked sugar chains.

TABLE II. Structural variation found in the outer chain moieties of complex type asparagine-linked sugar chains.

| I. | Gal$\beta$1→4GlcNAc$\beta$1→ |
|---|---|
| II. | NeuAc$\alpha$2→6Gal$\beta$1→4GlcNAc$\beta$1→ |
| III. | NeuAc$\alpha$2→3Gal$\beta$1→4GlcNAc$\beta$1→ |
| IV. | NeuAc$\alpha$2 |
| | ↓ |
| | 6 |
| | NeuAc$\alpha$2→3Gal$\beta$1→3GlcNAc$\beta$1→ |
| V. | NeuAc$\alpha$2 |
| | ↓ |
| | 6 |
| | NeuAc$\alpha$2→4Gal$\beta$1→3GlcNAc$\beta$1→ |
| VI. | Gal$\beta$1→4GlcNAc$\beta$1→ |
| | 3 |
| | ↑ |
| | Fuc$\alpha$1 |
| VII. | Gal$\beta$1→3Gal$\beta$1→4GlcNAc$\beta$1→ |
| VIII. | (Gal$\beta$1→4GlcNAc$\beta$1→3)$_n$•Gal$\beta$1→4GlcNAc$\beta$1→ |

$N$-acetylglucosamine residue of the core. Often, the outer chains are trisaccharides, Sia→Gal→GlcNAc, and the galactose-to-$N$-acetylgluco-samine linkage is mostly $\beta$1→4. However, several exceptional cases in which the linkage is $\beta$1→3 are found. In addition, the various kinds of outer chains listed in Table II have been reported up to the present.

Those which fall into the last subgroup were found by studying the larger sugar chains of hen egg albumin (16, 17). As shown in Fig. 1, these oligosaccharides have hybrid structures which include features of both high mannose-type and complex-type sugar chains. They have a common core of Man$\alpha$1→6(Man$\alpha$1→3)Man$\beta$1→4GlcNAc$\beta$1→4GlcNAc and an $N$-acetylglucosamine residue and/or a Gal$\beta$1→4GlcNAc group are linked to one $\alpha$-mannosyl residue as in the case of complex-type sugar chains while one or two $\alpha$-mannose residues are linked to another $\alpha$-mannosyl residue like high mannose-type sugar chains.

The constancy of the arrangement of the mannosyl residues in the core region of the three groups of asparagine-linked sugar chains may indicate that they are formed by a common biosynthetic processing pathway recently elucidated independently by Kornfeld, Robbins, and Summers (18–20) (Fig. 2). Whether all asparagine-linked sugar chains are formed by a series of processing pathways as shown in Fig. 2 is not clear, because several unusual asparagine-linked sugar chains have also been reported (Fig. 3).

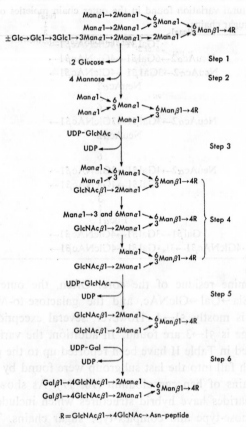

Fig. 2. Processing in the biosynthetic pathway of complex-type asparagine-linked sugar chains of glycoproteins.

## III. SPECIES-SPECIFIC STRUCTURAL DIFFERENCES IN THE OUTER CHAIN MOIETY OF COMPLEX-TYPE ASPARAGINE-LINKED SUGAR CHAINS

The species-related differences in the sugar chain moiety of glyco-proteins has been quite well documented recently. In Fig. 4, two typical examples of such cases are summarized.

Fibronectin isolated from bovine plasma contains four different asparagine-linked sugar chains (24), while that isolated from human plasma contains only two types of biantennary complex-type sugar

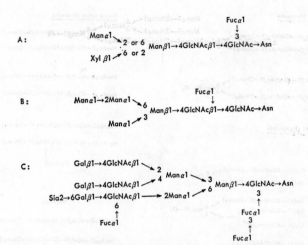

Fig. 3. Structures of unusual asparagine-linked sugar chains found in glycoproteins. A is from pineapple stem bromelain (21), B from lima bean lectin (22), and C from human secretory component (23).

chains (25). The sugar chains of rat and human $\alpha_1$-acid glycoproteins also showed quite large structural differences (26). First of all, none of the sugar chains from human $\alpha_1$-acid glycoprotein contains an $\alpha$-fucosyl residue on the proximal $N$-acetylglucosamine residue while some of them contain one or two $\alpha$-fucosyl residues in their outer chain moieties. The sugar chains of rat glycoprotein contain a Gal$\beta$1$\rightarrow$3GlcNAc group in their outer chain moiety, which was not found in those of human glycoprotein. Approximately 80% of the rat sugar chains were of bi-antennary complex type while more than 90% of the human sugar chains were tri- and tetraantennary. These distinct differences may arise from the different sets of glycosyltransferases included in the Golgi membrane of the two animals. As will be described in the next section, the structures of sugar chains synthesized are determined by the presence or absence of a series of glycosyltransferases.

## IV. PROBLEM OF MICROHETEROGENEITY OF THE SUGAR CHAINS OF GLYCOPROTEINS

After the discovery of glycogen synthetase by Leloir and Cardini

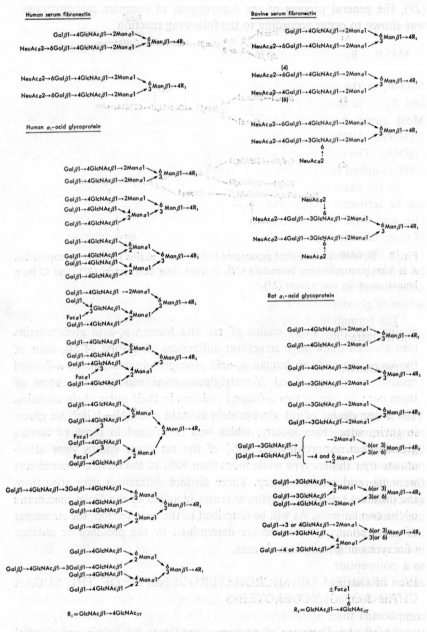

Fig. 4.   Species differences of asparagine-linked sugar chains of glycoproteins.

(27), the general process of the biosynthesis of complex carbohydrates was shown to occur according to the following reaction.

$$N\text{-}O\text{-}R_1 + R_2 \cdots \longrightarrow N\text{-}OH + R_1\text{-}R_2 \cdots$$

$N\text{-}O\text{-}R_1$ is the sugar nucleotide which works as a glycosyl ($R_1$) donor and $R_2 \cdots$ is the acceptor, which is mostly a part of the sugar chains. Most sugar nucleotides are nucleoside diphosphate sugars such as UDP-Glc, UDP-Gal, GDP-Man, GDP-Fuc, UDP-GlcNAc, and UDP-GalNAc. The only exception is the activated form of sialic acid, in which CMP is linked to one of the sialic acid residues.

In the cases of microorganisms and plants, a single monosaccharide can be activated by different nucleotides. For example, ADP-Glc, UDP-Glc, and GDP-Glc are formed in plants and are used for the synthesis of three different glucans: starch, cellulose, and callose, respectively.

In the case of mammals, however, each monosaccharide is activated by a single nucleotide (28). Therefore, all sugar chains of glycoconjugates found in mammalian tissues are formed solely by the concerted action of glycosyltransferases.

The formation of mucin-type sugar chains on the polypeptide backbone is started by the following reaction (29).

$$UDP\text{-}GalNAc + Ser(Thr)\text{-}peptides \longrightarrow$$
$$GalNAc\alpha 1 \rightarrow Ser(Thr)\text{-}peptides + UDP$$

The $N$-acetylgalactosaminyl transferase which catalyzes the reaction recognizes either serine or threonine residues in the polypeptide chains. Evidence that not all serine and threonine residues are glycosylated may indicate that the enzyme recognizes a part larger than a single serine or threonine residue in the polypeptide (30). The fact that purified $N$-acetylgalactosaminyl transferase can transfer $N$-acetylgalactosamine only to high molecular weight acceptors and not to serine and threonine supports this assumption. However, the nature of the minimum acceptor structure is not yet understood. Once an $N$-acetylgalactosamine residue is added to a polypeptide, sugar chain elongation starts step by step through the action of a series of glycosyltransferases.

The formation of asparagine-linked sugar chains is much more complicated than mucin-type sugar chains. It starts with the enzymatic transfer of a tetradecasaccharide from a lipid-bound intermediate (31) to

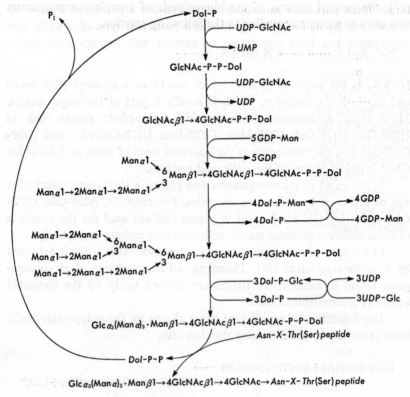

Fig. 5. Pathway of asparagine-linked sugar chain formation. Dol-P, dolicholphosphate; Dol-P-P, dolicholpyrophosphate.

the asparagine residue of the Asn-X-Ser (or Thr) sequence in polypeptide chains (Fig. 5) (32). In the tripeptide sequence, X can be any amino acid other than proline (33). The tetradecasaccharide is then subjected to a series of processing pathways to form Manα1→6(GlcNAcβ1→2Manα 1→3)Manβ1→4GlcNAcβ1→4GlcNAc→Asn as summarized in Fig. 2. The hexasaccharide chain is then converted to various complex-type sugar chains by elongation of the outer chain moieties. This elongation again proceeds by the concerted action of a series of glycosyltransferases. Since no template as in the case of protein biosynthesis is included in these sugar chain elongation reactions, the structures of the final sugar chains produced are determined by the specificity of each glycosyltransferase for a particular nucleotide sugar and for a particular glycose acceptor, and by its ability to synthesize a particular type of linkage. This mechanism has been considered to explain well the so-called microheterogeneity of sugar chains widely found in the carbohydrate moiety of glycoproteins. A shortage of certain nucleotide sugars, changes in relative glycosyltransferase levels, and many other factors can theoretically induce changes in the major sugar chain structures. Although the formation of the oligomannosyl core portion of asparagine-linked sugar chains is controlled as previously described, the processing pathway and outer chain elongation steps can produce partially completed sugar chains. In fact, the sugar chains of hen egg albumin (16, 17, 34) and bovine pancreatic ribonuclease (35) both have a single asparagine-linked sugar chain and were shown to be a mixture of a series of biosynthetic intermediates together with a completed sugar chain.

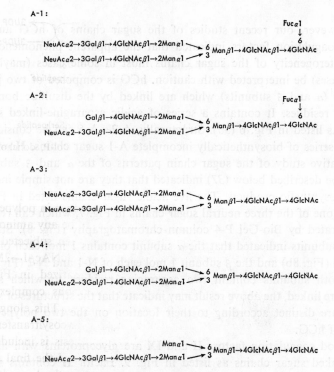

Fig. 6.  Structures of asparagine-linked sugar chains of hCG.

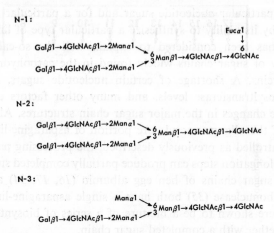

Fig. 7. Structures of the three neutral sugar chains obtained from the asparagine-linked sugar chains of hCG by sialidase treatment.

However, our recent studies of the sugar chains of hCG and of blood coagulation factors indicated that the apparent phenomenon of microheterogeneity of the sugar chains must in some cases (maybe in many cases) be interpreted with caution. hCG is composed of two poly-peptides ($\alpha$ and $\beta$ subunits) which are linked by the disulfide bond of cysteine residues. It contains a series of acidic asparagine-linked sugar chains as listed in Fig. 6 (36). Structurally, A-2∼A-5 can be considered to be a series of biosynthetically incomplete A-1 sugar chains. However, comparative study of the sugar chain patterns of the $\alpha$ and $\beta$ subunits as will be described below (37) indicated that they are not simple incomplete biosynthetic products. The five acidic sugar chains listed in Fig. 6 contain one of the three neutral sugar chains in Fig. 7, which can readily be separated by Bio-Gel P-4 column chromatography (Fig. 8). Studies of two subunits indicated that the $\alpha$ subunit contains 1 mol each of N-2 and N-3 (Fig. 8b) and the $\beta$ subunit 1 mol each of N-1 and N-2 (Fig. 8c). Since both subunits contain two asparagine residues to which sugar chains are linked, the above result may indicate that the structures of sugar chains are distinct according to their location on the two polypeptide chains of hCG.

Blood coagulation factors II and IX are glycoproteins with aspar-agine-linked sugar chains as listed in Fig. 9. Factor II contains sugar chains A, B, and C and factor IX contains the sialylated forms of D and

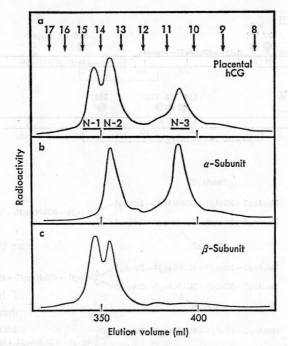

Fig. 8.    Gel filtration patterns of the desialylated oligosaccharide fractions. Radioactive oligosaccharides were passed through a Bio-Gel P-4 column and eluted with water. The *black arrows* indicate the eluting positions of glucose oligomers, and *numbers* indicate the glucose units.

E in addition. Amino acid sequence studies have revealed that factors II and IX have 3 and 4 asparagine residues with sugar chains, respectively. Comparative studies of the oligosaccharide patterns obtained from glycopeptide fragments of factor II and IX have revealed that the sugar chain C is linked only to Asn[376] of factor II and Asn[261] of factor IX. The triantennary sugar chains (D and E) were not linked to Asn[261] of factor IX.

The specific distribution of different sugar chains at different loci of glycoprotein molecules cannot be explained by current knowledge of the biosynthetic mechanism of the asparagine-linked sugar chains. However, from the viewpoint of the functional role of sugar chains in hCG, this fact is of particular interest because it indicates the possibility that each sugar chain in a glycoprotein molecule may be playing a different role, such as that of a recognition signal.

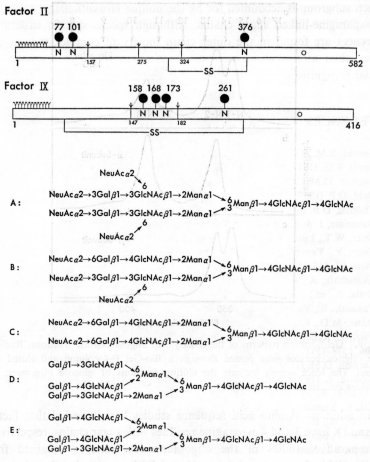

Fig. 9. Structures of bovine blood coagulation factors II and IX and their sugar chains (38).

## SUMMARY

Two types of sugar chains, the mucin-type and the asparagine-linked, are found in the cell surface glycoproteins. The asparagine-linked sugar chains are further classified into three subgroups: high mannose type, complex type and hybrid type. The occurrence of common core structure in each subgroup is accounted for by the unique biosynthetic pathway of the asparagine-linked sugar chains. Although species-specific structural differences are found in the outer chain moiety of complex type sugar chains, their structural variation suggests that they play key roles in cellular recognition.

## REFERENCES

1. Gesner, B.M. & Ginsburg, V. (1964) *Proc. Natl. Acad. Sci. U.S.* **52**, 750–756
2. Spiro, R.G. (1972) *Methods Enzymol.* **28**, 3–43
3. Carlson, D.M. (1967) *Anal. Biochem.* **20**, 195–198
4. Bahl, O.P. (1969) *J. Biol. Chem.* **244**, 567–575
5. Thomas, D.B. & Winzler, R.J. (1969) *J. Biol. Chem.* **244**, 5943–5946
6. Baenziger, J. & Kornfeld, S. (1974) *J. Biol. Chem.* **249**, 7270–7281
7. Shier, W.T., Lin, Y., & De Vries, A.L. (1975) *FEBS Lett.* **54**, 135–138
8. Endo, Y., Yamashita, K., Han, Y.N., Iwanaga, S., & Kobata, A. (1977) *J. Biochem.* **82**, 545–550
9. Gottschalk, A., Bhargava, A.S., & Murty, V.L.N. (1972) in *Glycoproteins* (Gottschalk, A., ed.) pp. 810–829, Elsevier, Amsterdam
10. Takasaki, S., Yamashita, K., & Kobata, A. (1978) *J. Biol. Chem.* **253**, 6086–6091
11. Oates, M.D., Rosbottom, A.C., & Schrager, J. (1974) *Carbohydr. Res.* **34**, 115–137
12. Feizi, T., Kabat, E.A., Vicari, G., Anderson, B., & Marsh, W.L. (1971) *J. Immunol.* **106**, 1578–1592
13. Carlson, D.M. (1968) *J. Biol. Chem.* **243**, 616–626
14. Kobata, A. (1979) *Anal. Biochem.* **100**, 1–14
15. Takasaki, S., Mizuochi, T., & Kobata, A. (1982) *Methods Enzymol.* **83**, 263–268
16. Tai, T., Yamashita, K., Ito, S., & Kobata, A. (1977) *J. Biol. Chem.* **252**, 6687–6694
17. Yamashita, K., Tachibana, Y., & Kobata, A. (1978) *J. Biol. Chem.* **253**, 3862–3869
18. Tabas, I., Schlesinger, S., & Kornfeld, S. (1978) *J. Biol. Chem.* **253**, 716–722
19. Turco, S.J., Stetrom, B., & Robbins, P.W. (1977) *Proc. Natl. Acad. Sci. U.S.* **74**, 4411–4414
20. Hunt, L.A., Etchison, J.R., & Summers, D.F. (1978) *Proc. Natl. Acad. Sci. U.S.* **75**, 754–758
21. Fukuda, M., Kondo, T., & Osawa, T. (1976) *J. Biochem.* **80**, 1223–1232
22. Misaki, A. & Goldstein, I.J. (1977) *J. Biol. Chem.* **252**, 6995–6999
23. Purkayastha, S., Rao, C.V.N., & Lamm, M.E. (1979) *J. Biol. Chem.* **254**, 6583–6587
24. Takasaki, S., Yamashita, K., Suzuki, K., Iwanaga, S., & Kobata, A. (1979) *J. Biol. Chem.* **254**, 8548–8553
25. Takasaki, S., Yamashita, K., Suzuki, K., & Kobata, A. (1980) *J. Biochem.* **88**, 1587–1594
26. Yoshima, H., Matsumoto, A., Mizuochi, T., Kawasaki, T., & Kobata, A. (1981) *J. Biol. Chem.* **256**, 8476–8484
27. Leloir, L.F. & Cardini, C.E. (1957) *J. Am. Chem. Soc.* **79**, 6340–6341

28. Ginsburg, V. & Kobata, A. (1971) in *Structure and Function of Biological Membranes* (Rothfield, L.I., ed.) pp. 439–459, Academic Press, New York
29. McGuire, E.J. & Roseman, S. (1967) *J. Biol. Chem.* **242**, 3745–3747
30. Hill, H.D., Jr., Schwyzer, M., Steinman, H.M., & Hill, R.L. (1977) *J. Biol. Chem.* **252**, 3799–3804
31. Li, E., Tabas, I., & Kornfeld, S. (1978) *J. Biol. Chem.* **253**, 7762–7770
32. Marshall, R.D. (1974) *Biochem. Soc. Symp.* **40**, 17
33. Bause, E. (1979) in *Glycoconjugates* (Schauer, R., Boer, P., Buddecke, E., Kramer, M.F., Vliegenthart, J.F.G., & Wiegandt, H., eds.) pp. 228–229, Georg Thieme, Stuttgart
34. Tai, T., Yamashita, K., Ogata-A., M., Koide, N., Muramatsu, T., Iwashita, S., Inoue, Y., & Kobata, A. (1975) *J. Biol. Chem.* **250**, 8569–8575
35. Liang, C.-J., Yamashita, K., & Kobata, A. (1980) *J. Biochem.* **88**, 51–58
36. Endo, Y., Yamashita, K., Tachibana, Y., Tojo, S., & Kobata, A. (1979) *J. Biochem.* **85**, 669–679
37. Mizuochi, T. & Kobata, A. (1980) *Biochem. Biophys. Res. Commun.* **97**, 772–778
38. Mizuochi, T., Fujikawa, K., Titani, K., & Kobata, A. (1981) in *Glycoconjugates* (Yamakawa, T., Osawa, T., & Handa, S., eds.) pp. 267–268, Japan Sci. Soc. Press, Tokyo

# Structure and Function of Outer Membrane of *Escherichia coli* : A Reconstitution Study

SHOJI MIZUSHIMA

*Laboratory of Microbiology, Faculty of Agriculture,*
*Nagoya University, Nagoya 464, Japan*

Based on the cell surface structure, bacteria can be taxonomically divided into two groups, Gram-positives and Gram-negatives. As shown in Fig. 1, a cell of Gram-positive bacteria is surrounded by a thick peptidoglycan layer (cell wall) and the cytoplasmic membrane, while that of Gram-negative bacteria is surrounded by three layers. They are, from outside to inside, the outer membrane, peptidoglycan layer, and cytoplasmic membrane. Therefore, the outer membrane is an organella that is specific to Gram-negative bacteria. The outer membrane is rather simple in protein composition, having a smaller variety of proteins in very large quantities. In addition, many of the proteins have an unusually stable structure. These made it easy to purify and characterize these proteins. Furthermore, they made it possible to study the interaction between these proteins and other outer membrane constituents such as lipopolysaccharide, which is turn led us to the reconstitution work.

One of the powerful techniques for studying the structure and function of a subcellular organella is that of reconstitution. This involves dissociation of an organella into its molecular components, purification of the individual components, and reconstitution of an organella whose structure and function is the same as that of the original organella. The

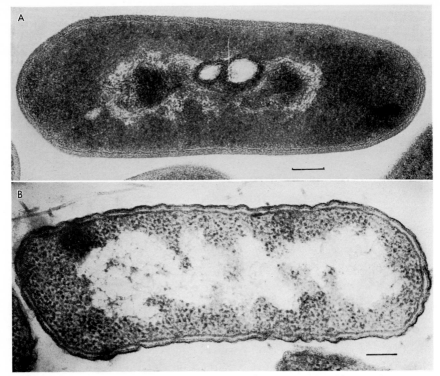

Fig. 1. Thin sectioning electron micrographs of Gram-positive (A: *Brevibacterium lipolyticum*; a photograph taken by Dr. A. Hirata) and Gram-negative (B: *E. coli*) bacteria. Bars represent 100 nm.

reconstitution technique has been successfully applied to structure-function analyses of many organelles such as ribosomes, flagella, and viruses.

This chapter deals with the structure and function of the outer membrane, mainly of *Escherichia coli*. Research in this field has developed quite rapidly in the last few years. However, it is not intended in this chapter to cover all of the aspects. Instead, special attention is given to the use of the reconstitution technique. More detailed discussion of the structure and function in general and other related topics will be found in the volume of outer membrane edited by Inouye (*1*) and other recent reviews (*2, 3*).

## I.   COMPONENTS OF OUTER MEMBRANE

The method to separate the outer membrane and the cytoplasmic membrane was first developed by Miura and Mizushima (4, 5) and is schematically presented in Fig. 2. Several modifications including that by Osborn *et al.* (6) have been made on this method. In all cases, the outer membrane is separated from the cytoplasmic membrane because of its greater density due to the presence of lipopolysaccharide (LPS).

The outer membrane thus isolated is composed of protein, phospholipid, and LPS. LPS is a molecule unique to the outer membrane; its chemical structure and a closed packing model are given in Figs. 3 and 4, respectively. It contains the hydrophobic fatty acid region and the hydrophilic polysaccharide region that includes 2-keto-3-deoxyoctonate (3-deoxy-D-mannooctulosonate), a characteristic component of the LPS molecule. The content of phospholipid in the outer membrane is much less than that in the cytoplasmic membrane. The outer membrane contains a few species of so-called "major proteins." These proteins have been purified and studied in many laboratories with different names. They are listed in Table I together with names the same as those of the

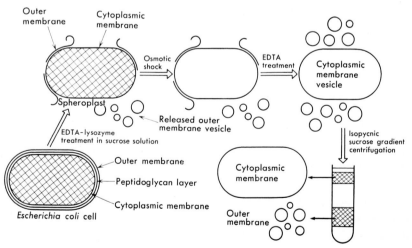

Fig. 2.   Separation of outer membrane and cytoplasmic membrane of *E. coli*: Schematic representation.

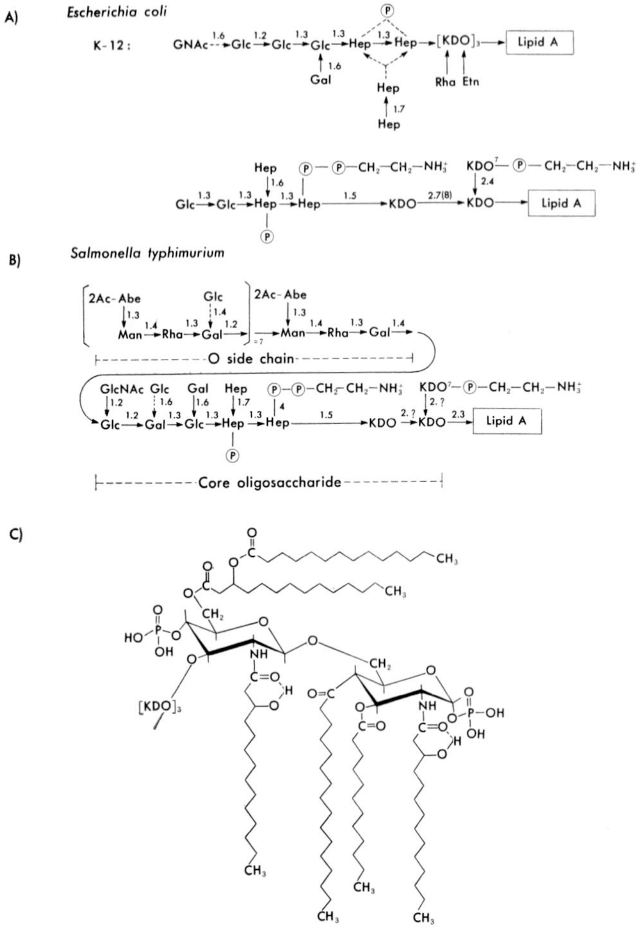

Fig. 3.   Structure of LPS. A) *E. coli.* B) *Salmonella typhimurium.* C) A proposed structure of lipid A. Glc, glucose; GNAc, *N*-acetylglucosamine; Gal, galactose, Hep,
heptose; KDO, 2-keto-3-deoxyoctonate; Rha, rhamnose; 2Ac-Abe, 2-acetylabecose;
Man, mannose.

structural genes. Since those named after structural genes became the
most popular, these will be used in this chapter for the proteins except
the smallest one that has been called the lipoprotein. It should be noted
that the definition of major protein is rather arbitrary. A minor protein
may become a major protein when its production is fully induced.

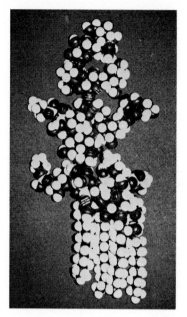
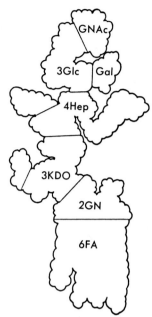

Fig. 4.  Closed packing model of LPS of *E. coli* K-12 and its sketch. FA, fatty acid; GN, glucosamine; KDO, 2-keto-3-deoxyoctonate; Hep, heptose; Glc, glucose; Gal, galactose; GNAc, *N*-acetylglucosamine.

TABLE I.  Nomenclature of major outer membrane proteins of *E. coli* (2).

| Protein nomenclature | | | | Recomended nomenclature |
|---|---|---|---|---|
| Mizushima | Henning | Schnaitman | Lugtenberg | |
| O-8 | Ib | lb | c | OmpC |
| O-9 | Ia | la | b | OmpF |
| O-10 | II* | 3a | d | OmpA |
| O-18 | IV | | | Major lipoprotein |

## II.  PROPERTIES OF MAJOR OUTER MEMBRANE PROTEINS

### 1.  The Lipoprotein

This protein was first discovered as a form covalently bound to the peptidoglycan layer (7) (Fig. 5). Later the protein was found to exist as a free form as well, about 1/3 as the bound form and the rest as the free form (8). Both forms were found in the outer membrane fraction when the outer membrane was separated from the cytoplasmic membrane

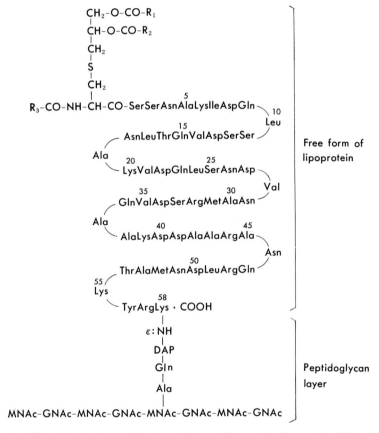

Fig. 5.  Chemical structure of the bound form of the major lipoprotein of the outer membrane. DAP, diaminopimelic acid; MNAc, *N*-acetylmuramic acid; GNAc, *N*-acetylglucosamine. $R_1$, $R_2$, and $R_3$ represent hydrocarbon chains of fatty acids.

after lysozyme digestion of the peptidoglycan layer (*9, 10*). This protein is the most abundant protein in the cell in terms of numbers of molecules. Recently several new lipoproteins were discovered in *E. coli* (*11*). These are suggested to contain the fatty acid-bearing glycerylcysteine residue at the N-terminus, this indicates that the structure of these new lipoproteins resembles that of the major lipoprotein to some extent. However, their molecular weights are much larger and the nunbers of molecules per cell are much smaller than those of the major lipoprotein.

## 2.   OmpF and OmpC

These two proteins are found in *E. coli* K12, while only OmpF is found in *E. coli* B (*12*). The two proteins resemble each other in many biochemical and physicochemical aspects and are thought to be products of genes that have developed from a single ancestral gene (13). They are characterized by their tight, but noncovalent, association with the peptidoglycan layer (*14, 15*), trimer structure (*16–18*), high content of β-structure (*19*), and resistance to detergents including sodium dodecyl sulfate (SDS) (*19*). On the other hand, the biosynthesis of the two proteins is affected in opposite directions by medium osmolarity. The high osmolarity is inductive for OmpC and suppressive for OmpF (*20, 21*).

Another interesting characteristic of these proteins is that they form passive diffusion pores that enable small hydrophilic substances to pass through the outer membrane (*3*). Because of this property, these proteins are called "porin" as well.

## 3.   OmpA

Like OmpF/OmpC, this protein is also SDS-resistant and has a high β-structure content (*19*). Although the physiological importance of this major protein is still unclear, it is reported to be required in F-pilus mediated conjugation (*22–24*).

## III.   RECONSTITUTION OF OUTER MEMBRANE

The SDS-resistant nature of major outer membrane proteins made it possible to carry out a reconstitution of the outer membrane from individual components dissolved in SDS solution (*25, 26*). Summary of the reconstitution work to be discussed here is schematically presented in Fig. 6. When either OmpF or OmpC was mixed with LPS in SDS solution and dialyzed against a solution containing magnesium ions to remove SDS and to introduce magnesium ions, membrane vesicles with an ordered hexagonal lattice structure were formed (Fig. 7A). A lattice constant is about 7 nm. Both the protein and LPS are essential for the lattice formation. In the presence of the lipoprotein-bearing peptidoglycan sacculus, the lattice formation predominantly takes place on the entire surface of the sacculus (Fig. 7B). The bound form of the lipoprotein is essential for the lattice formation on the sacculus. When the lipoprotein-free peptidoglycan sacculus was used, no lattice structure

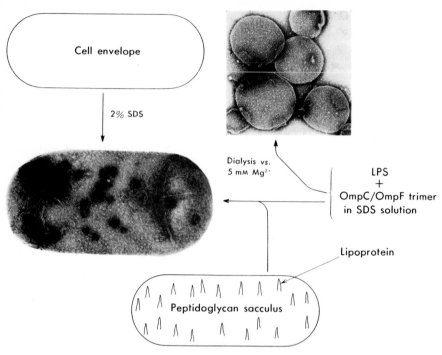

Fig. 6.    Schematic representation of reconstitution of outer membrane of *E. coli*.

was formed on the sacculus; instead, vesicles with a lattice structure, which were essentially the same as those formed from OmpF/OmpC and LPS, were formed independent of the sacculus. The lattice structure thus formed on the prptidoglycan sacculus was essentially the same as that observed in the cell envelope when it was treated with SDS (*14, 27*). Importance of the bound form of the major lipoprotein in the interaction between the outer membrane and the peptidoglycan layer was also shown under the same conditions using a lipoprotein-negative mutant (*26*). When the cell envelope of this mutant was treated with SDS solution, no lattice structure was observed on the peptidoglycan sacculus. Instead, most of it was observed rather independent of the peptidoglycan layer as the reconstituted sample with the lipoprotein-free peptidoglycan shows.

Since the reconstitution system described above includes most of the major constituents of the cell envelope (the outer membrane plus the

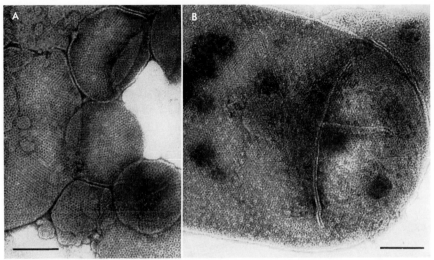

Fig. 7. Electron micrographs of reconstituted *E. coli* outer membrane (negative staining). A) Reconstitution with OmpC and LPS. B) Reconstitution with OmpC, LPS, and lipoprotein-bearing peptidoglycan sacculus. Bars represent 100 nm.

peptidoglycan layer) and since the structure formed is essentially the same as that of the SDS-treated cell envelope, it is highly likely that the reconstituted one represents a basal interaction that occurs in the *E. coli* cell envelope. However, it should be stressed that such an ordered lattice structure may not exist in the intact cell envelope. The orderliness may be distorted by the presence of other outer membrane constituents such as OmpA or phospholipids that may also interact with molecules used for the reconstitution.

The reconstitution work has provided a couple of interesting pieces of information. First of all, it shows that at least as far as the outer membrane is concerned specific interactions between constituents determine the assembly of the biomembrane, as in cases of assembly of other organelles such as ribosomes, flagella, and viruses. It further shows that the peptidoglycan layer most likely plays an important role as a scaffold in the outer membrane assembly. It was thought that the outer membrane and the peptidoglycan layer were independent organelles. Based on this, they have been isolated from each other and characterized rather separately. However, evidence has accumulated that the structure and assembly of these organelles are related. To achieve a balanced duplication of a bacterial cell, the biogenesis of cellular organelles must somehow influence each other. Interaction between cellular organelles in terms of biogenesis, structure and function should be studied more extensively to understand the cell structure as the unit of life.

The reconstitution work also provided us with useful tools for studying the structure and function of the cell envelope. These are discussed in the following two sections focusing on two subjects; one is the interaction between membrane proteins and LPS and the other is the roles of cell surface components in bacteriophage infection.

## IV. INTERACTION BETWEEN OUTER MEMBRANE PROTEINS AND LPS

Interaction between protein and lipid is one of the most important forces in biomembranes. From a structural point of view, this interaction is a basic force in maintaining the two-dimensional membrane structure. From a functional point of view, evidence has accumulated that the function of many membrane proteins is controlled by the presence of the phospholipid or LPS. Little is known, however, about the nature of interaction. This is partly due to the lack of a proper system to work with.

The fact that outer membrane proteins OmpF/OmpC and LPS assemble into an ordered hexagonal lattice has been successfully used for

TABLE II. Lattice constants of reconstituted hexagonal lattice structure. Reconstitutions were carried out from the OmpC trimer and LPS or its derivatives with the ratios given in the first column. Electron micrographs of negatively stained specimens were subjected to optical diffractometry and lattice constants were determined. The ratio of OmpC trimer to LPS (or its derivatives) are expressed as molar equivalents; one LPS molecule was regarded as being equivalent to six fatty acids (from Ref. 28).

| OmpC trimer/LPS or its derivatives | Lattice constant (nm) of samples reconstituted with | | | |
|---|---|---|---|---|
| | LPS | Heptoseless LPS | Lipid A | Fatty acid from LPS |
| 1:1–2 | $6.5\pm0.1$ | $6.5\pm0.1$ | $6.6\pm0.1$ | $6.5\pm0.1$ |
| 1:3 | $6.7\pm0.3$ | $6.6\pm0.1$ | — | $6.5\pm0.1$ |
| 1:6 | — | — | $6.6\pm0.1$ | — |
| 1:10 | $7.5\pm0.3$ | $7.2\pm0.3$ | — | $6.5\pm0.1$ |
| 1:20 | — | — | $6.8\pm0.2$ | — |
| 1:30 | — | — | $7.1\pm0.2$ | $6.6\pm0.1$ |

studying the interaction between these molecules (28). LPS and its derivatives were used to determine the moiety of the LPS molecules responsible for the formation of the lattice structure. A summary of the study with OmpC is presented in Table II. The minimum amount of LPS required to form a hexagonal lattice structure with OmpF/OmpC is probably one molecule per OmpF/OmpC trimer. Since the molecular weight of these trimers is about 110,000 and that of LPS is only 4,300, it is highly likely that trimers are the main constituent of the lattice and LPS is required to maintain the hexagonal arrangement of the trimers. However, as far as the hexagonal arrangement of the OmpC protein is concerned, the function of LPS can be taken over by an equivalent amount of fatty acid. Since one LPS molecule contains six fatty acid residues, it is possible that one LPS molecule interacts almost equally with three subunits belonging to three different trimers. Each LPS molecule contains three $\beta$-hydroxymyristic acid residues, a fatty acid specific to LPS. However, it is uncertain whether this fatty acid residue plays a specific role in the arrangement of these trimers.

Although one LPS molecule, or even the fatty acid moiety derived from it, for each trimer is enough to form a hexagonal lattice, the spacing of the lattice thus formed is appreciably smaller than that observed in the cell envelope which has been treated with SDS solution. The hexagonal lattice with a lattice constant almost the same as that of the SDS-treated envelope is obtained when a 10-fold amount of LPS is used. It should be noted that the increase of the lattice constant is not observed with the equivalent amount of lipid A or fatty acid. This fact suggests that the carbohydrate moiety of LPS induces a conformation change of the trimer, which in turn results in the increase of the lattice constant. The heptoseless LPS results in a partial increase of the lattice constant. This suggests that both the distal end of the polysaccharide region, including both heptose and the 2-keto-3-deoxyoctonate regions, is involved in the interaction. The importance of the latter region in the cell surface has already been suggested (29). The importance of the heptose region in the outer membrane assembly has also been suggested, since the heptoseless mutation has often resulted in an appreciable decrease of the outer membrane protein content (30–32).

Essentially the same results were obtained with LamB, the receptor protein for phage $\lambda$, although the lattice constant provided with LamB was slightly larger than that with OmpF/OmpC (33). This is probably due to the fact that the molecular weight of LamB is larger than that of OmpF/OmpC.

The interaction between LPS and OmpF/OmpC has also been suggested from a functional point of view. The receptor activity for bacteriophages T4 and TuIb depends on both LPS and OmpC in the K-12 strain (34–36, 38) and that for phages T2 and TuIa requires both OmpF and LPS (37, 38). Furthermore, the presence of LPS is essential for OmpF functioning as a channel for small hydrophilic molecules (39).

## V.  ROLES OF CELL SURFACE COMPONENTS OF *E. COLI* K-12 IN BACTERIOPHAGE INFECTION

The infection process of a bacteriophage begins with adsorption of phage particles to a specific receptor on the host cell surface. T-even type bacteriophages, phages having a complicated structure, undergo a series of morphological changes upon adsorption to *E. coli* cells. The adsorption process may be divided into the following stages: i) binding of the distal end of individual tail fibers to the receptor on the host cell surface, ii) pinning down of tail pins to the cell surface, iii) contraction of the tail sheath and insertion of the core into the cell, and iv) injection of phage DNA into the host bacterium. The roles of individual constituents of phages in the infection process have been studied rather extensively, while a larger part of the roles of cell surface components remains to be studied. This was due to the lack of a proper system to work with.

Table III is a summary of the phages having T-even type structure and the receptor components of *E. coli* K-12 which are recognized by

TABLE III.   Receptor components for *E. coli* K-12 phages having T-even type structure.

| Phage | Receptor components | | References |
|-------|---------|-------|-----------|
| T2 | OmpF | LPS | (37) |
| T4 | OmpC | LPS | (34, 35) |
| T6 | Tsx | LPS[a] | (37) |
| TuIa | OmpF | LPS | (38) |
| TuIb | OmpC | LPS | (36, 38) |
| TuII* | OmpA | LPS | (38) |

[a] Yamato, I., personal communication.

them. In all cases two components were found to constitute individual receptors. One is an outer membrane protein which is specific for individual phages, and the other is LPS which is common to them all. Since OmpF/OmpC and LPS can be incorporated into the reconstituted cell surface discussed above, the phage infection process can be studied with it.

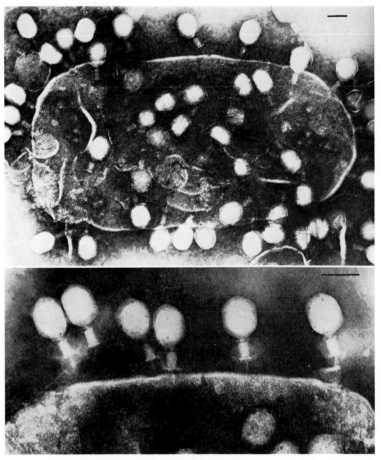

Fig. 8.   Negative stain of phage T4 adsorbed to the *E. coli* cell surface reconstituted from OmpC, LPS, and lipoprotein-bearing peptidoglycan sacculus. Bars represent 100 nm. (Furukawa, H., Yamada, H., & Mizushima, S. (1979) *J. Bacteriol.* **140**, 1071–1080)

    The cell surface reconstituted from OmpC, LPS, and the lipoprotein-
bearing peptidoglycan sacculus is active as a receptor for phage T4 up
to the stage of core insertion through the cell surface (Fig. 8) (40). The
core insertion is clearly observed with a thin sectioned specimen. However, the ejection of phage DNA does not take place. Since the peptidoglycan layer is not directly involved in the interaction with long tail
fibers and tail pins (40), the results indicate that receptors for both these
parts consist of either or both LPS and OmpC. Evidence has accumulated
that in T-even phages long tail fibers seem to play an important role in
the specific attachment of the phage to sensitive bacteria (41, 42). Therefore, one can speculate that individual proteins listed in Table III (OmpC
for phage T4) may be involved in the specific interaction with the long
tail fibers while LPS, which is the component common to all of the
receptors, may act as a receptor for the short tail fibers. However, this
does not exclude the involvement of LPS in the interaction with the long
tail fibers. This molecule is also assumed to constitute a part of the
receptor for the long tail fibers either directly or indirectly through an
interaction with OmpC (25, 28). In E. coli B, LPS is assumed to be the
sole receptor component for both the long tail fibers and the tail pins
(43).

    The OmpC-LPS lattice assembled on the surface of the lipoprotein-
bearing peptidoglycan layer is much more active as a receptor than the
lattice alone, suggesting that the peptidoglycan layer plays some role in
the infection process. This layer probably allows OmpC and LPS to
assemble widely onto its surface so that the OmpC-LPS complex can
interact with individual distal ends of the long tail fibers on a single phage
particle (40). It may also act as a rigid support to the otherwise flexible
outer membrane so that the core can penetrate this membrane upon tail
sheath contraction. It was observed that TuIb, a T4-like phage that was
adsorbed and contracted on the reconstituted cell surface, cannot thrust
its core completely through the surface in the absence of the peptidoglycan layer. Instead, the core extruded from the tail sheath pushed
down the reconstituted surface, thus making a small indentation on the
surface (F. Yu, H. Yamada, and S. Mizushima, unpublished observation).

    Electron microscopic studies also revealed that the tip of a core of
phages adsorbed on the reconstituted cell surface penetrates through the
surface, indicating that no additional component is required for the

core penetration (40). Together with the finding that the peptidoglycan layer supports the core penetration, it is therefore most probable that the core penetration is simply carried by the mechanical force of the contraction of the tail sheath which was held on the cell surface by both tail pins and long tail fibers.

The process of DNA ejection from phage particles was studied using T4 particles artificially contracted either by a chemical method or by introducing heat-sensitive mutations. Phospholipids, a major component of the cytoplasmic membrane, induce ejection of phage DNA from the contracted phage (44). Both phosphatidylglycerol and cardiolipin were found to be active components, while phosphatidylethanolamine was inactive (45). Experiments with phospholipid liposomes revealed that the DNA ejection was induced by the interaction of the core tip of the contracted phage with the phospholipid bilayer (Fig. 9).

The liposome-induced DNA ejection was also observed with the reconstituted cell surface (45). When the reconstituted cell surface having a closed sacculus structure was fragmented and mixed with liposomes, some fraction of the liposomes was able to get inside the sacculus. Under

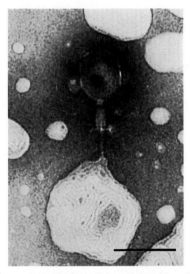

Fig. 9.   Interaction of a contracted phage T4 particle with a liposome. The wild type T4 was contracted by urea-treatment and mixed with liposomes prepared from *E. coli* phospholipids. Bar represents 100 nm (negative staining). (Furukawa, H. & Mizu-shima, S. (1982) *J. Bacteriol.* **150**, in press)

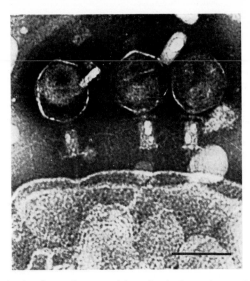

Fig. 10.   DNA ejection from phage particles adsorbed on a reconstituted cell surface. The cell surface reconstituted from OmpC, LPS, and lipoprotein-bearing peptidoglycan sacculus was fragmented and mixed with the wild type phage T4. It was then further incubated with liposomes and examined after negative staining. Bar represents 100 nm. (Furukawa, H. & Mizushima, S. (1982) *J. Bacteriol.* **150**, in press)

these conditions, DNA in a phage that was adsorbed on the cell surface was ejected upon its interaction with liposomes of the core tip that had penetrated the cell surface (Fig. 10). As a control experiment, the reconstituted cell surface having an intact sacculus structure was used without fragmentation. Although this intact sacculus caused the same morphological changes of T4 as the fragmented one did, the addition of liposomes to the phage-sacculus complex did not cause DNA ejection. The results strongly indicate that only liposomes that have reached the core tips from the inside of the reconstituted cell surface are able to indiuce DNA ejection; neither a membrane potential nor a pH gradient across the membrane is required for this action. It should be noted, however, that phage DNA that has been ejected is not taken up by liposomes. This indicates that an additional factor(s) is required for DNA injection across the cytoplasmic membrane into the cytoplasm; that factor may be a membrane protein. Alternatively, it may be a membrane potential, since this potential is required for DNA injection into intact cells upon phage infection (*46*).

## SUMMARY

Reconstitution techniques have been successfully applied to structure-function analyses of many organelles, including ribosomes, flagella, and viruses. Experimental data presented in this chapter shows that a reconstitution technique is also applicable to the study of the outer membrane. This means that the basic structure of the outer membrane on the peptidoglycan layer is also determined by specific interactions between major constituents as in the case of other organelles. However, it is uncertain whether we can extend such an idea to the molecular assembly of biomembranes in general. The reconstitution system was also quite useful for the study of the mechanism of bacteriophage infection, that is, the roles of individual cell surface components in the bacteriophage infection process.

### Acknowledgment

Work carried out in our laboratory was supported by grants from the Ministry of Education, Science and Culture of Japan and the Toray Science Foundation.

## REFERENCES

1. Inouye, M. (ed.) (1979) *Bacterial Outer Membranes: Biogenesis and Functions*, Wiley-Interscience, New York.
2. Osborn, M.J. & Wu, H.C.P. (1980) *Annu. Rev. Microbiol.* **34**, 369–422
3. Nikaido, H. & Nakae, T. (1979) in *Advances in Microbial Physiology* (Rose, A.H. & Tempest, D.W., eds.) Vol. 20, pp. 163–250, Academic Press, New York
4. Miura, T. & Mizushima, S. (1968) *Biochim. Biophys. Acta* **150**, 159–161
5. Miura, T. & Mizushima, S. (1969) *Biochim. Biophys. Acta* **193**, 268–276
6. Osborn, M.J., Gander, J.E., Parisi, E., & Carson, J. (1972) *J. Biol. Chem.* **247**, 3962–3972
7. Braun, V. (1975) *Biochim. Biophys. Acta* **415**, 335–377
8. Inouye, M., Shaw, J., & Shen, C. (1972) *J. Biol. Chem.* **247**, 8154–8159
9. Braun, V. & Bosch, V. (1973) *FEBS Lett.* **34**, 302–306
10. Lee, N. & Inouye, M. (1974) *FEBS Lett.* **39**, 167–170
11. Ichihara, S., Hussain, M., & Mizushima, S. (1981) *J. Biol. Chem.* **256**, 3125–3129
12. Ichihara, S. & Mizushima, S. (1977) *J. Biochem.* **81**, 1525–1530
13. Ichihara, S. & Mizushima, S. (1978) *J. Biochem.* **83**, 1095–1100
14. Rosenbusch, J.P. (1974) *J. Biol. Chem.* **249**, 8019–8029

15. Hasegawa, Y., Yamada, H., & Mizushima, S. (1976) *J. Biochem.* **80**, 1401–1409
16. Palva, E.T. & Randall, L.L. (1978) *J. Bacteriol.* **133**, 279–286
17. Yu, F., Ichihara, S., & Mizushima, S. (1979) *FEBS Lett.* **100**, 71–74
18. Nakae, T., Ishii, J., & Tokunaga, M. (1979) *J. Biol. Chem.* **254**, 1457–1461
19. Nakamura, K. & Mizushima, S. (1976) *J. Biochem.* **80**, 1411–1422
20. van Alphen, W. & Lugtenberg, B. (1977) *J. Bacteriol.* **131**, 623–630
21. Kawaji, H., Mizuno, T., & Mizushima, S. (1979) *J. Bacteriol.* **140**, 843–847
22. Manning, P.A., Puspurs, A., & Reeves, P. (1976) *J. Bacteriol.* **127**, 1080–1084
23. Schweizer, M. & Henning, U. (1977) *J. Bacteriol.* **129**, 1651–1652
24. Havekes, B.J., Lugtenberg, B.J.J., & Hoekstra, W.P.M. (1976) *Mol. Gen. Genet.* **146**, 43–50
25. Yamada, H. & Mizushima, S. (1978) *J. Bacteriol.* **135**, 1024–1031
26. Yamada, H. & Mizushima, S. (1981) *Agric. Biol. Chem.* **45**, 2083–2090
27. Steven, A.C., Heggeler, B.T., Muller, R., Kistler, J., & Rosenbusch, J.P. (1977) *J. Cell Biol.* **72**, 292–301
28. Yamada, H. & Mizushima, S. (1980) *Eur. J. Biochem.* **103**, 209–218
29. Rick, P.D., Fung, L.W., Ho, C., & Osborn, M.J. (1977) *J. Biol. Chem.* **252**, 4904–4912
30. Koplow, J. & Goldfine, H. (1974) *J. Bacteriol.* **117**, 527–543
31. Ames, G.F., Spudick, E.N., & Nikaido, H. (1974) *J. Bacteriol.* **117**, 409–416
32. Lugtenberg, B., Peters, R., Bernheimer, H., & Berendsen, W. (1976) *Mol. Gen. Genet.* **147**, 251–262
33. Yamada, H., Nogami, T., & Mizushima, S. (1981) *J. Bacteriol.* **147**, 660–669
34. Mutoh, N., Furukawa, H., & Mizushima, S. (1978) *J. Bacteriol.* **136**, 693–699
35. Henning, U. & Jann, K. (1979) *J. Bacteriol.* **137**, 664–666
36. Yu, F., Yamada, H., & Mizushima, S. (1981) *J. Bacteriol.* **148**, 712–715
37. Hantke, K. (1978) *Mol. Gen. Genet.* **164**, 131–135
38. Datta, D.B., Arden, B., & Henning, U. (1977) *J. Bacteriol.* **131**, 821–829
39. Schindler, H. & Rosenbusch, J.P. (1981) *Proc. Natl. Acad. Sci. U.S.* **78**, 2302–2306
40. Furukawa, H., Yamada, H., & Mizushima, S. (1979) *J. Bacteriol.* **140**, 1071–1080
41. Beckendorf, S.K., Kim, J.S., & Lielausis, I. (1973) *J. Mol. Biol.* **73**, 17–35
42. Crawford, J.T. & Goldberg, E.B. (1980) *J. Mol. Biol.* **139**, 679–690
43. Wilson, J.H., Luftig, R.B., & Wood, W.B. (1970) *J. Mol. Biol.* **51**, 423–434
44. Baumann, L., Benz, W.C., Wright, A., & Goldberg, E.B. (1970) *Virology* **41**, 356–364
45. Furukawa, H. & Mizushima, S. (1982) *J. Bacteriol.* **150**, in press.
46. Labedan, B. & Goldberg, E.B. (1979) *Proc. Natl. Acad. Sci. U.S.* **76**, 4669–4673
47. Manning, P.A. & Reeves, P. (1978) *Mol. Gen. Genet.* **158**, 279–286

# Synthesis and Translocation of Mitochondrial Matrix Enzymes in Higher Animals with Special Reference to δ-Aminolevulinate Synthase

GORO KIKUCHI AND NORIO HAYASHI

*Department of Biochemistry, Tohoku University School of Medicine, Sendai 980, Japan*

While the mitochondrion contains its own genetic system, about 90% of the mitochondrial proteins are coded for by the nuclear genome (*1*). These proteins include proteins in the intermembrane space, some hydrophobic proteins which localize mostly in the inner membrane, and presumably all the proteins in the mitochondrial matrix. These proteins are translocated from the site of synthesis to the site of their function across the outer membrane or both the outer and inner membranes of mitochondria. Synthesis of the mitochondrial proteins coded for by the nuclear genome, as well as their integration into mitochondria, have been extensively studied in yeast and *Neurospora* (*2–4*), but those studies with fungi are concerned mainly with proteins such as cytochromes, ATPase, ADP/ATP carrier protein, and cytochrome *c* peroxidase, which are integrated or attached to the inner membrane or localize in the intermembrane space, and citrate synthesis (*5*) is the only soluble matrix protein that has been extensively studied.

On the other hand, papers dealing with the mitochondrial matrix proteins of higher animals have been accumulating recently. The matrix enzymes investigated include glutamate dehydrogenase (rat liver) (*6–8*),

131

malate dehydrogenase (rat liver (*9, 10*), rabbit liver (*11*)), carbamoyl phosphate synthetase I (rat liver (*12–17*), bullfrog liver (*18*)), ornithine transcarbamoylase (rat liver) (*16, 17, 19–25*), alanine aminotransferase (rat liver (*26, 27*), chicken heart (*28*)), serine aminotransferase (rat liver) (*29, 30*), ornithine aminotransferase (rat liver) (*31*), fumarase (rat liver) (*32*), glycerate kinase (rat liver) (*33, 34*), phosphoenolpyruvate carboxykinase (bullfrog liver (*35–38*), chicken liver (*39, 40*)), adrenodoxin and adrenodoxin reductase (bovine adrenal cortex) (*41*), and δ-aminolevulinate (ALA) synthase (rat liver (*42–49*), chicken liver (*50*), chick embryo liver (*51–53*)). Studies in cell-free protein synthesis systems have indicated that most, if not all, of the mRNA directing the synthesis of mitochondrial matrix proteins is confined to free polysomes, and the mitochondrial proteins synthesized on cytoribosomes are translocated into mitochondria by a post-translational mechanism. Also, precursors of several mitochondrial enzymes have been demonstrated in the cytosol fraction, although enzyme proteins are not always synthesized as larger molecules with additional peptide sequences. The general features of the synthesis and intracellular translocation of the mitochondrial matrix proteins in higher animals appear to be essentially similar to those observed for the mitochondrial proteins in yeast and *Neurospora crassa*.

Among these animal matrix proteins, ALA synthase is unique in several aspects. ALA synthase is the rate-limiting and regulatory enzyme in the heme biosynthetic pathway (*54*). ALA synthase in the liver can be easily induced by the administration of porphyrinogenic drugs such as allylisopropylacetamide (AIA) and 3,5-dicarbethoxy-1,4-dihydrocollidine (DDC), thus giving rise to overproduction of ALA and some other intermediates of porphyrin biosynthesis (*54, 55*). In animals with chemically-induced hepatic porphyria, ALA synthase accumulates to a considerable degree not only in mitochondria but also in the cytosol fraction of the liver (*42, 50, 56–60*) although the extent of the enzyme accumulation in the cytosol fraction is variable according to the species of animals and drugs used; the accumulation is particularly significant when the enzyme is induced in rats by the administration of AIA, but the enzyme does not appreciably accumulate in the liver cytosol fraction in chick embryo (*51, 61*). Moreover, the transfer of ALA synthase from the cytosol into the mitochondria is inhibited by heme, the end-product of the biosynthetic pathway (*43, 45, 48, 62*). To our knowledge ALA synthase is the only example for which the occurrence of regulation of intracellular translocation has been clearly demonstrated. This inhibition represents a new mechanism of the feedback regulation of metabolism in the sense that the inhibition of intracellular translocation of ALA synthase would bring about a reduction in the activity of porphyrin biosynthesis. In addition, the synthesis of ALA synthase is strongly suppressed by heme possibly at both the transcriptional and post-transcriptional levels (*48, 55, 62, 63*).

In the present paper we will review the synthesis and intracellular translocation of various mitochondrial matrix enzymes in higher animals with particular attention paid to studies made in Japan and with special reference to ALA synthase.

## I. SYNTHESIS AND INTRACELLULAR TRANSLOCATION OF ALA SYNTHASE

### 1. Induction of Synthesis of ALA Synthase in Rat Liver

The accumulation of ALA synthase in the liver cytosol fraction has been observed in the rat (*42, 56, 58, 59*), mouse (*57, 60*), guinea pig (*59*), and chicken (*50*), but not in the chick embryo (*51, 61*), after the administration of AIA or DDC. An example of enzyme induction in AIA-treated rats is shown in Fig. 1. In an earlier stage of induction, the level of ALA synthase in the mitochondrial fraction was considerably higher than the level in the cytosol fraction, suggesting that the enzyme synthesized on cytoribosomes is rapidly incorporated into mitochondria where the enzyme functions physiologically. In fact, ALA synthase disappeared from the liver cytosol with a half-life of about 20 min when animals were injected with cycloheximide, whereas the apparent half-life of ALA synthase in the mitochondria was 60 to 70 min under comparable experimental conditions (*42, 47*).

In a later stage of induction, that is, after the second administration of AIA (Fig. 1), the degree of increase in the mitochondrial enzyme was relatively small, while the enzyme level in the cytosol fraction increased steadily, so that the enzyme level in the cytosol far exceeded the enzyme level in the mitochondria. ALA synthase may accumulate in the extramitochondrial space when the rate of synthesis of the enzyme exceeds the rate of its transfer into the mitochondria; the capacity of mitochondria to accommodate the enzyme seems to be limited to a certain level.

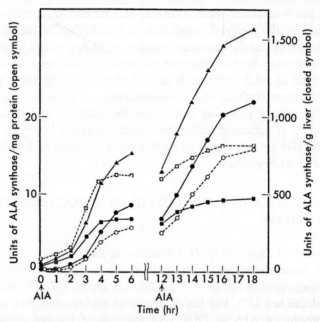

Fig. 1. Time courses of AIA-induced increase in ALA synthase in mitochondria and the cytosol of rat liver. Rats were injected subcutaneously with AIA (300 mg/kg body weight) at the times indicated by the arrows. □, ■ mitochondrial fraction; ○, ● cytosol fraction; ▲ total activity.

## 2. Inhibition of Intracellular Translocation of ALA Synthase by Heme

Inhibition by heme of the intracellular transfer of ALA synthase was first found in 1970 in our laboratory (43, 64). As shown in Fig. 2, when hemin was injected into rats which had been treated with AIA, the level of mitochondrial ALA synthase decreased sharply, and in turn, the enzyme level in the cytosol fraction increased considerably. The total activity in the whole liver was therefore not appreciably decreased by hemin administration. If heme inhibits the translocation of ALA synthase from the cytosol into mitochondria, it is natural that the enzyme level in mitochondria decreases rapidly after the hemin injection, since the apparent half-life of the mitochondrial enzyme is as short as 60 to 70 min (42, 47, 65). In addition, it was found that the apparent disappearance time of cytosolic ALA synthase was markedly elongated by the injection of hemin (43, 47). As shown in Fig. 3, when both cycloheximide and

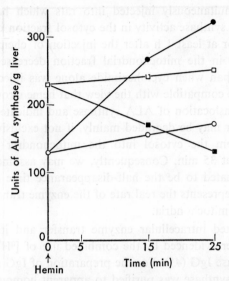

Fig. 2. Effect of hemin on ALA synthase activity in mitochondria and the cytosol of AIA-treated rat liver. A single dose (300 mg/kg body weight) of AIA was administered subcutaneously to rats; 3.5 h after AIA administration, hemin (5 mg/kg body weight) was administered intravenously. □ mitochondria, AIA alone; ■ mitochondria, AIA and hemin; ○ cytosol, AIA alone; ● cytosol, AIA and hemin.

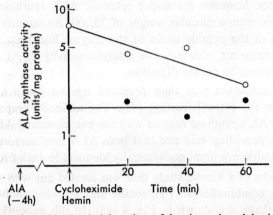

Fig. 3. Effect of combined administration of hemin and cycloheximide on ALA synthase activity in the liver of AIA-treated rats. Hemin (5 mg/kg body weight) and cycloheximide were simultaneously injected into rats 4 h after AIA. ● cytosol; ○ mitochondria.

hemin were simultaneously injected into rats which had been treated with AIA, ALA synthase activity in the cytosol fraction did not appreciably decrease for at least 1 h after the injection of chemicals, while the enzyme activity in the mitochondrial fraction decreased more rapidly ($t_{1/2} = 35$ min) than when cycloheximide alone was injected (47). These observations are compatible with the view that heme strongly inhibits the intracellular translocation of ALA synthase and indicate that ALA synthase in rat liver may be degraded mainly, if not exclusively, after being translocated from the cytosol into the mitochondria with an actual half-life of about 35 min. Consequently, we may assume that the value of 20 min estimated to be the half-disappearance time of the cytosolic ALA synthase represents the real rate of the enzyme transfer *in vivo* from the cytosol into mitochondria.

The suspected intracellular enzyme transfer and its regulation by heme was further evidenced by the combined use of [$^3$H]leucine and an anti-ALA synthase IgG (45). For the preparation of IgG specific to ALA synthase, ALA synthase was purified to apparent homogeneity from the liver cytosol fraction of AIA-treated rats by employing several purification steps including papain digestion (44). The specific activity of the preparation finally obtained was 73,000 units/mg protein. The minimum molecular weight of the purified ALA synthase was 51,000 as judged by sodium dodecyl sulfate (SDS)-polyacrylamide gel electrophoresis. As will be described later, however, the native cytosolic ALA synthase was found to have a minimum molecular weight of 75,000. Apparently a considerable portion of the peptide chain of the enzyme had been removed by the papain treatment, although the catalytic activity of ALA synthases was not decreased by papain digestion.

Rabbit antiserum was then prepared against the ALA synthase purified from the cytosol fraction (44). The antibody prepared against the cytosolic ALA synthase reacted with the mitochondrial ALA synthase equally well, providing evidence that both ALA synthases in the cytosol and mitochondria are immunochemically identical to each other.

The results of a kinetic study that was carried out with AIA-treated rats, using a combination of [$^3$H]leucine and a rabbit antibody specific to ALA synthase, are shown in Fig. 4 (45). In this experiment, all rats were treated with AIA for 3.5 h, then the rats were divided into two groups, and one group of rats was injected with hemin. Five min later, all rats were injected with [$^3$H]leucine. In rats not treated with hemin,

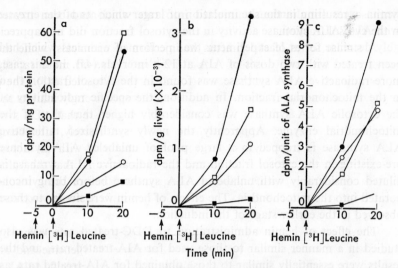

Fig. 4.   Incorporation of [³H]leucine into ALA synthase in mitochondria and the cytosol of rat liver and effect of hemin on it in an earlier stage of induction. Rats had been treated with AIA for 3.5 h. Five min after hemin (5 mg/kg body weight), each rat was injected with L-[4,5-³H]leucine (100 $\mu$Ci/100 g body weight). Radioactivity of ALA synthase in mitochondria and the cytosol of the liver was determined after quantitative immunoprecipitation. Symbols are the same as in Fig. 2.

more radioactive ALA synthase was found in the mitochondrial fraction than in the cytosol fraction when compared on the basis of either mg protein or g liver. This is in agreement with the observation shown in Fig. 1 that in the early stage of induction, the enzyme activity in the mitochondrial fraction increased faster than that in the cytosol fraction. However, when compared in terms of the specific radioactivity of ALA synthase (the radioactivity of immunoprecipitated protein per unit of ALA synthase), the value for the enzyme in both fractions was quite similar. This agrees with the idea that the ALA synthase newly synthesized on cytoribosomes is rapidly incorporated into the mitochondria.

In the hemin-treated rat, the incorporation of [³H]leucine into the mitochondrial ALA synthase was greatly reduced, while the incorporation of radioactivity into the cytosolic ALA synthase was markedly increased. It is particularly significant that the specific radioactivity of the mitochondrial ALA synthase was greatly reduced in hemin-injected rats. This is a good indication that hemin actually inhibited the translocation of ALA

synthase, resulting in the accumulation of larger amuonts of the enzyme
in the cytosol fraction.

A similar series of experiments was performed with rats which had
been treated with two doses of AIA at 12-h intervals (45). In this case,
more radioactive ALA synthase was found in the cytosol fraction than
in the mitochondrial fraction. In addition, the specific radioactivity of
the cytosolic ALA synthase was considerably higher than that of the
mitochondrial enzyme. Apparently the newly synthesized radioactive
ALA synthase is trapped in a large pool of unlabeled ALA synthase
pre-existing in the cytosol fraction and the radioactive ALA synthase is
diluted considerably with unlabeled ALA synthase before being incorporated into the mitochondria. The effects of hemin were similar to those
observed in the earlier stage of the induction.

The effect of hemin administration to DDC-treated rats was also
studied in a manner similar to those used for AIA-treated rats, and the
results were essentially similar to those obtained for AIA-treated rats, as

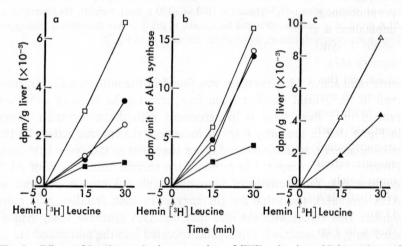

Fig. 5. Effects of hemin on the incorporation of [³H]leucine into ALA synthase of
the liver of DDC-treated rats. All rats were given two doses of DDC at 10-h intervals
and hemin was injected intravenously into one group of rats 5 h after the 2nd dose of
DDC. Five min after hemin, each rat was injected with [³H]leucine (180 μCi/100 g
body weight) interperitoneally. □ mitochondria, DDC alone; ■ mitochondria,
DDC and hemin; ○ cytosol, DDC alone; ● cytosol, DDC and hemin; △ mito-
chondria+cytosol, DDC alone; ▲ mitochondria+cytosol, DDC and hemin.

can be seen from Fig. 5 (*48*). In DDC-treated rats, however, the incorporation of radioactivity into the cytosolic enzyme was not appreciably increased in hemin-injected rats as compared with that in hemin-uninjected control rats. Therefore, as shown in Fig. 5c, the total incorporation of [³H]leucine into hepatic ALA synthase in hemin-injected rats was about half of that in hemin-uninjected rats. This suggests that heme considerably inhibits the synthesis of ALA synthase in DDC-treated rats. Possibly a post-transcriptional event is involved in this inhibition since the incorporation of [³H]leucine into ALA synthase was suppressed immediately after the administration of hemin, while the half-life of mRNA for ALA synthase in the rat liver has been estimated to be 65–75 min (*65, 66*).

## 3. Inhibition of ALA Synthase Synthesis by Heme

There have been several lines of evidence that heme inhibits the synthesis of ALA synthase at either the transcriptional level or a post-transcriptional level, or both (*54, 55*). Particularly important is the finding that in acute intermittent porphyria the increased excretion of porphobilinogen is associated with a combination of two enzymatic anomalies: a genetically determined primary deficiency of uroporphyrinogen I synthetase and a secondary rise of hepatic ALA synthase activity (*67*). The latter possibly results from derepression of ALA synthase and this would in turn generate an increased supply of heme precursors to overcome the partial block in heme biosynthesis.

With respect to the site of action of heme in the inhibition of ALA synthase synthesis, several investigators (*61, 68–70*) proposed the post-transcriptional inhibition on the bases of the observations that the enzyme activity in cultured chick embryo liver cells began to decrease immediately after the addition of hemin into the culture medium at a rate that was higher than those observed after the addition of actinomycin D and was comparable to those obtained by cycloheximide. The data shown in Fig. 5c suggest that a similar situation also occurs in DDC-treated rats. Consistent with this view, we have observed that when hemin was injected into DDC-treated rats, the increase in the accumulation of the cytosolic enzyme after the administration of hemin was relatively small, whereas the enzyme level in the mitochondria rapidly decreased immediately after the hemin injection and therefore the enzyme level in the whole liver also rapidly decreased (Fig. 6) (*48*). On the other

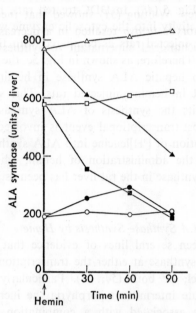

Fig. 6. Effects of hemin on ALA synthase activity in the liver of DDC-treated rats. Rats were given two doses of DDC (300 mg/kg body weight) at 10-h intervals. Hemin (5 mg/kg body weight) was injected intravenously into one group of rats 5 h after the 2nd dose of DDC. Symbols are the same as in Fig. 5.

hand, the level of ALA synthase did not appreciably decrease for at least the initial 30 min when either $\alpha$-amanitin or actinomycin D was administered to DDC-treated rats (48). These observations support the view that a post-transcriptional event(s) is involved in the inhibition by heme in DDC-treated rats.

It should be noted, however, that, as can be seen from Fig. 4, incorporation of [³H]leucine into ALA synthase in AIA-treated rats was not appreciably inhibited by the administration of the same dose of hemin (5 mg/kg body weight) as used for DDC-treated rats during the experimental period. In AIA-treated rats, the machinery of ALA synthase synthesis may have been made less sensitive to heme, especially at the post-transcriptional level, for some unknown reason. In fact, when larger doses of hemin were given to AIA-treated rats, the enzyme activity in the whole liver decreased considerably (49).

There is evidence that heme might also inhibit enzyme synthesis at a transcriptional step. Whiting (*51*) showed that the administration of hemin to chick embryos *in ovo* resulted in a decrease in the ability of the liver post-mitochondrial supernatant to synthesize the enzyme *in vitro*, and suggested that heme acted to decrease the amount of mRNA for the enzyme. Srivastava *et al*. (*71*) also presented evidence for a transcriptional inhibition of ALA synthase synthesis by heme in the chick embryo liver.

Recently we have observed that the administration of hemin to AIA-treated rats caused a time-dependent and dose-dependent decrease in the ability of liver polysomes to synthesize ALA synthase *in vitro* in the reticulocyte lysate system (*63*). Polysomes isolated from the liver of AIA-treated rats could actively synthesize ALA synthase *in vitro*, while the activity of polysomes isolated from the liver of control rats was negligibly small (Fig. 7). When hemin was administered to AIA-treated rats, however, the synthetic activity of liver polysomes was reduced by about 50% at 60 min after the hemin administration. This value is com-

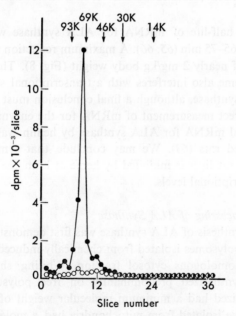

Fig. 7. Electrophoretic analysis of ALA synthase synthesized in a reticulocyte lysate system with polysomes from the liver of AIA-induced or uninduced rats. ● AIA-induced rats; ○ uninduced rats.

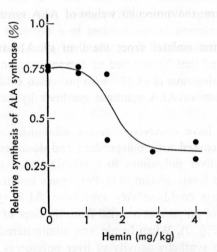

Fig. 8. Extents of decrease in the activity of liver polysomes to synthesize ALA synthase *in vitro* in AIA-treated rats 60 min after the administration of various doses of hemin. Rats had been pretreated with AIA for 3.5 h before hemin administration.

parable to the half-life of mRNA for ALA synthase which has been estimated to be 65–75 min (*65, 66*). A maximum reduction was achieved at a hemin dose of nearly 2 mg/kg body weight (Fig. 8). These data would indicate that heme also interferes with a transcriptional step of the synthesis of ALA synthase, although a final conclusion must await confirmation by the direct measurement of mRNA for the enzyme. A reduction in the functional mRNA for ALA synthase by hemin was also observed for DDC-treated rats (*63*). We may conclude that synthesis of ALA synthase in the rat liver is inhibited by heme at both the transcriptional and post-transcriptional levels.

### 4. Possible Processing of ALA Synthase

Cell-free synthesis of ALA synthase was first demonstrated by Whiting (*51*) using polysomes isolated from chemically-induced chick embryo livers and a homologous cytosol fraction. Whiting showed that the enzyme was synthesized predominantly on free polysomes, and the enzyme synthesized had a minimum molecular weight of about 70,000, while the enzyme isolated from mitochondria had a molecular weight of about 49,000. Subsequently Brooker *et al.* (*52*) isolated poly(A) RNA from the liver of chick embryo and translated this poly(A) RNA in a wheat germ cell-free system; the molecular weight of ALA synthase synthesized was again 70,000.

Using polysomes isolated from the liver of AIA-treated rats and using a reticulocyte lysate system, we also demonstrated that free polysomes were the major site of ALA synthase synthesis (46). The ALA synthase synthesized *in vitro*, when examined by SDS-gel electrophoresis followed by fluorography, showed a molecular weight about 2,000 larger than that of the cytosolic ALA synthase which was estimated to have a minimum molecular weight of 75,000 (to be published). On the other

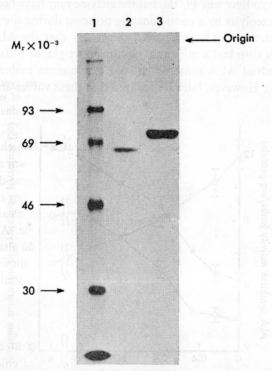

Fig. 9. Comparison of molecular weights of cytosolic and mitochondrial ALA synthases of rat liver. Cytosolic and mitochondrial ALA synthases were labeled *in vivo* by injection of [$^{14}$C]leucine and [$^{3}$H]leucine, respectively, into AIA-treated rats. Labeled ALA synthases were collected by immunoprecipitation from the cytosol fraction and the sonic extract of mitochondria and analyzed by SDS-gel electrophoresis followed by fluorography. Lane 1, marker proteins; lane 2, mitochondrial ALA synthase; lane 3, cytosolic ALA synthase.

hand, the size of ALA synthase isolated from the mitochondria was considerably smaller than that of the cytosolic ALA synthase, as can be seen from Fig. 9; its minimum molecular weight was about 66,000. The value of 66,000 is practically equal to the value of 65,000 which was recently reported by Ades and Harpe (53) for a mature form of embryonic chick liver ALA synthase. ALA synthase may be synthesized as a precursor with a molecular weight of about 77,000, then is processed first to the cytosolic form of the enzyme with a molecular weight of 75,000 which is then incorporated into the mitochondrial matrix, being accompanied by processing. Whiting (51) reported that the mitochondrial ALA synthase of chick embryo liver was 49,000, but the enzyme may have been subjected to limited proteolysis by a contaminating protease during the preparation of the enzyme. Also we have reported previously that the ALA synthase synthesized *in vitro* had a minimum molecular weight of 51,000, whereas the mitochondrial ALA synthase showed a minimum molecular weight of 45,000 (46). However, later we realized that these values are wrong.

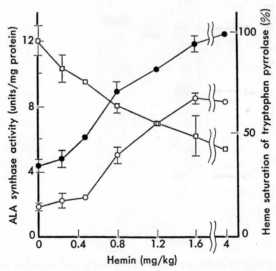

Fig. 10.  Effects of administration of various doses of hemin on ALA synthase activity in mitochondria and the cytosol and on the degree of heme saturation of tryptophan pyrrolase in the liver of AIA-treated rats. Rats were given indicated doses of hemin intravenously 3 h after the administration of AIA and were killed 30 min later for enzyme assay. □ ALA synthase in mitochondria; ○ ALA synthase in the cytosol; ● heme saturation of tryptophan pyrrolase.

## 5.  *Physiological Significance of the Regulatory Effects of Heme on ALA Synthase*

Although it is clear now that both the synthesis of ALA synthase and its transfer from the cytosol into mitochondria are inhibited by the administration of hemin, a question remains as to whether the suspected regulation by heme could function under physiological conditions. To answer this question, we examined comparatively the effects of hemin administration on both the behavior of ALA synthase and the degree of heme saturation of tryptophan pyrrolase (49), taking advantage of the fact that tryptophan pyrrolase has been assumed to be a very sensitive marker for assessing subtle changes in the liver heme concentration (72, 73).

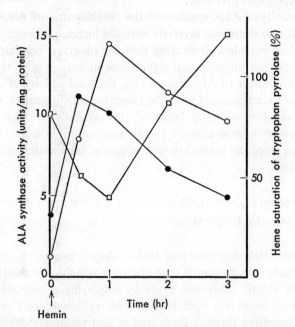

Fig. 11.   Time courses of changes in ALA synthase activity in mitochondria and the cytosol and of changes in the degree of heme saturation of tryptophan pyrrolase in the liver of AIA-treated rats. Rats were given hemin (1.6 mg/kg body weight) 3 h after AIA and were killed at the times indicated in the figure. Symbols are the same as in Fig. 10.

As shown in Fig. 10, the degree of heme saturation of tryptophan pyrrolase increased with the increasing hemin dosage, and the enzyme was almost fully saturated with heme at a dose of 1.6 mg/kg. Showing a very good correlation with the increase in heme saturation of tryptophan pyrrolase, ALA synthase activity in the mitochondria decreased and the enzyme activity in the cytosol increased, and the effects of hemin on both ALA synthase activities appeared to be nearly maximal at a dose of 1.6 mg/kg. Figure 11 shows the time course of the effects of hemin at a dose of 1.6 mg/kg. The intracellular distribution of ALA synthase changed significantly in response to the variation in the intracellular concentration of heme as estimated from the degree of heme saturation of tryptophan pyrrolase (49). These findings suggest that the intracellular translocation of ALA synthase can be affected by intracellular heme at concentrations lower than those necessary to fully saturate tryptophan pyrrolase.

The activity of ALA synthase in the mitochondria of AIA-untreated normal rats also decreased inversely with the increase in heme saturation of tryptophan pyrrolase, indicating that the observed regulatory action of heme really has physiological significance in normal rats (49). In normal rats, in contrast to AIA-treated rats, the total activity of ALA synthase in the liver decreased with the increased administration of hemin. The observed inhibition by heme of the synthesis of ALA synthase should involve a post-transcriptional mechanism, similar to DDC-treated rats, since the inhibition was apparent immediately after hemin administration.

## II. SYNTHESIS AND INTRACELLULAR TRANSLOCATION OF OTHER MATRIX ENZYMES

### 1. Glutamate Dehydrogenase and Malate Dehydrogenase

Glutamate dehydrogenase in the rat liver was the first matrix enzyme for which a kinetic study was made by employing a well-characterized antibody and an amino acid labeled with radioisotope. Godinot and Lardy (6) suggested through these studies that glutamate dehydrogenase was synthesized on polysomes associated with the endoplasmic reticulum and then transferred into the mitochondria. The malate dehydrogenase isozyme of the mitochondrial type was also reported to be synthesized by microsomes (9, 11).

These problems were reinvestigated later in more detail by Kawajiri *et al.* (*7, 8*) who came to a similar conclusion; they found greater amounts of [³H]leucine-labeled and immunoprecipitable nascent peptides on membrane-bound ribosomes than on free polysomes. Kawajiri *et al.* also reported the existence of particular light microsomal vesicles on the cytoplasmic surface of which both glutamate dehydrogenase and malate dehydrogenase (mitochondrial-type) were concentrated (*8*) The validity of these views, however, has been questioned (*4, 74*). Recently Aziz *et al.* (*10*) reported that malate dehydrogenase of rat liver mitochondria appeared to be synthesized on free polysomes as a precursor which is 1,500–2,000 daltons larger than the mature enzyme (34,500 daltons).

Mihara *et al.* (*75*) also re-examined the problems and found that both glutamate dehydrogenase and the mitochondrial malate dehydrogenase are synthesized on free polysomes as larger precursors; the precursor enzymes of glutamate dehydrogenase and malate dehydrogenase showed the molecular weights of 60,000 and 38,000, respectively, whereas those of the corresponding mature enzymes were 54,000 and 37,000, respectively. Morphological associations between the mitochondria and the endoplasmic reticulum (*76, 77*) or cytoplasmic ribosomes (*78*) have been reported, but there is no convincing evidence as yet that this association is related to the biosynthesis and intracellular translocation of mitochondrial proteins (*2, 3, 77, 79, 80*).

## 2. Carbamoyl Phosphate Synthetase and Ornithine Transcarbamoylase

These two enzymes catalyze the first two steps of urea biosynthesis in the liver. Carbamoyl phosphate synthetase is a dimer of identical subunits, each having a molecular weight of about 160,000. Shore and his associates demonstrated that the subunit of rat liver carbamoyl phosphate synthetase was synthesized by free polysomes as a precursor that is about 5,500 daltons larger than the mature subunit (*12*); the newly synthesized precursor was very rapidly transported into mitochondria in liver explants ($t_{1/2}$ is approximately 2 min) (*13, 14*). The putative precursor must have been subjected to processing either coincident with or immediately after its transmembrane uptake by the organelle. Mori *et al.* (*15, 16*) also reported similar findings for carbamoyl phosphate synthetase. Mori *et al.* (*15*) further demonstrated that a mitochondrial membrane preparation of rat liver could convert the precursor synthesized *in vitro* into an apparently mature form of the enzyme. On the other hand, bullfrog and tadpole carbamoyl phosphate synthetase that was synthesized *in vitro* was indistinguishable from the mature enzyme by any analytical methods used, indicating that the enzyme is not grossly modified during its transport into mitochondria in the frog (*18*).

Conboy *et al.* (*19*) and Mori *et al.* (*22*) showed that ornithine transcarbamoylase of the rat liver was synthesized in a reticulocyte lysate cell-free system as a precursor which was 3,400–4,000 daltons larger than the mature subunit (35,300–39,600 daltons). Both groups also showed that the cell-free product of ornithine transcarbamoylase was transported into isolated mitochondria and this was associated with a post-translational processing (*20–22*). Mori *et al.* (*23*) further showed that the proteolytic processing of ornithine transcarbamoylase would proceed in two steps through an intermediate with a size of about 37,000 daltons; a neutral protease, recovered mainly in a mitochondrial matrix fraction, was supposed to catalyze the first step (*23, 81*). They also found that mitochondria from rat non-hepatic tissues which lack ornithine transcarbamoylase, could take up and process the precursor to the mature form, suggesting that the transport-processing system is common to at least several matrix proteins (*24, 82*). Mori *et al.* revealed with isolated hepatocytes that the pulse-labeled precursor of ornithine carbamoyltransferase as well as that of carbamoyl phosphate synthetase appeared in the cytosol and then disappeared rapidly from the cytosol with apparent half-lives of about 1 to 2 min (*16*).

### 3.  *Aspartate Aminotransferase*

There exist two isozymes of aspartate aminotransferase, cytosolic and mitochondrial. Sonderegger *et al.* (*28*) showed that the mitochondrial aspartate aminotransferase synthesized in a cell-free system with polysomal mRNA from chicken heart was about 3,000 daltons larger than the mature subunit (44,500 daltons). Wada and his associates also showed that the mitochondrial aspartate aminotransferase synthesized in a reticulocyte lysate system with free polysomes from rat liver had a subunit which was about 2,000 daltons larger than the mature subunit (45,000 daltons) (*26*). The precursor was converted to an apparently mature form when incubated with a mitochondrial membrane preparation of rat liver (*26*). The $NH_2$-terminal amino acid of the enzyme newly synthesized *in vitro* was methionine (*26*) while the $NH_2$-terminal of mature mitochondrial aspartate aminotransferase of rat liver was serine (83), indicating that the precursor had an extra sequence attached to the $NH_2$-terminal end of the mature enzyme. Wada and associates (27) further demonstrated that the putative precursor of the mitochondrial enzyme was post-translationally incorporated into isolated mitochondria, being processed into the mature form. Mature mitochondrial asparate aminotransferase did not compete with the precursor in the transfer. This observation suggests that the precursor of the mitochondrial enzyme is transported into its destination much more readily than its mature form and is inconsistent with the report of Marra et al. (84–86) which indicated that rat liver mitochondria took up mature mitochondrial aspartate aminotransferase, though not its cytosolic isozyme.

### 4.  Serine Aminotransferase

Ichiyama and his colleagues synthesized this enzyme in a reticulocyte lysate system using a total RNA fraction prepared from the liver of glucagon-treated rats and found that this enzyme is synthesized as a precursor about 2,000 daltons larger than the mature peptide (40,000 daltons) (29). Serine aminotransferase which is immunochemically indistinguishable from the mitochondrial enzyme is also present in rat liver peroxisomes. However, they presented data suggesting that when synthesis of serine aminotransferase was induced by glucagon, the enzyme would be almost selectively incorporated into mitochondria but not into peroxisomes (30).

### 5.  Ornithine Aminotransferase

Hayashi et al. (31) synthesized rat liver ornithine aminotransferase in a reticulocyte lysate system using poly(A)mRNA from either free or bound polysomes and showed that the enzyme was preferentially synthesized on free ribosomes, but as one with the same size as the authentic purified enzyme (45,000 daltons).

### 6.  Fumarase

This enzyme distributes in both the cytosol and the mitochondria in the liver. Tuboi and his associates (32) purified and crystallized separately two fumarases located in the cytosolic and mitochondrial fractions of rat liver, respectively, and found that the two enzymes were indistinguishable from each other in molecular weight, the amino acid composition, kinetic constants, or immunochemical reactivity. This suggests the possibility that both enzymes are the products of the same gene; the mechanism by which the enzymes are distributed into two different subcellular compartments remains to be elucidated.

## 7. Glycerate Kinase

Sugimoto and his associates found that glycerate kinase, an enzyme which is responsible to gluconeogenesis from serine *via* hydroxypurvate, occurs in both the cytosol and mitochondria of rat liver (*33*), and reported that the two enzymes had identical kinetic and physical properties (*33*) and were immunologically indistinguishable (*34*). They suggested that both cytosolic and mitochondrial glycerate kinases arise from a common translation product and that dietary protein regulates the distribution of glycerate kinase to the cytosol and mitochondria.

## 8. Phosphoenolpyruvate Carboxykinase

This enzyme is supposed to be a key enzyme of gluconeogenesis and is found in both the cytosolic and the mitochondrial compartments of the liver, although the distribution of the enzyme in the subcellular compartments varies considerably according to the species of animal. Shukuya and his associates (*35*, *36*) showed that phosphoenolpyruvate carboxykinases in the cytosolic and mitochondrial fractions of the bull-frog liver were very similar in kinetic properties and in molecular weight, and were immunologically identical, while the two enzymes were different in p*I*. In tadpoles, only the amount of mitochondrial enzyme increases in response to thyroid hormone (*37*, *38*). They further showed that both the mitochondrial and cytosolic phosphoenolpyruvate carboxykinases were incorporated into intact mitochondria *in vitro* at the same rate (*37*). Kinetic studies with [³H]leucine in tadpoles support the idea that both enzymes in the cytosol and mitochondria of bullfrog liver are genetically the same protein (*37*).

On the other hand, Kochi and Kikuchi (*39*, *40*) reported the occurrence of two immunologically distinguishable phosphoenolpyruvate carboxykinases in the cytosol and mitochondria of chicken liver, respectively, although the cytosol-type enzyme was hardly detectable unless it was induced by the administration of hydrocortisone or isoproterenol to chickens. They demonstrated that the mitochondria-type enzyme was synthesized mainly on free polysomes as a putative precursor 3,000 to

4,000 daltons larger than the mature form of the mitochondria-type enzyme which had a molecular weight of 72,000 (39). The cytosol fraction of chicken liver also contained a considerable amount of phosphoenol-pyruvate carboxykinase which had an immunoreactivity identical to that of the mitochondria-type enzyme, but major portions of the enzyme seemed to be accounted for by release of the mitochondrial enzyme. Kinetic studies in vivo using [³H]leucine and antibody, however, revealed that some portions of phosphoenolpyruvate carboxykinase appearing in the liver cytosol would correspond to a precursor in transit to the mitochondria (40).

## 9.  Adrenodoxin and Adrenodoxin Reductase

Omura and his associates (41) studied the in vitro synthesis of adreno-doxin and adrenodoxin reductase by using free and bound polysomes isolated from the bovine adrenal cortex. The results indicated that adrenodoxin reductase was synthesized only by free polysomes and possibly as the mature-sized product. In contrast, adrenodoxin was synthesized by both free and bound polysomes as a precursor of 20,000 daltons, which was processed to the mature size of adrenodoxin (12,000 daltons) by in vitro incubation with adrenal cortex mitochondria.

## III.  COMMENTS

Recent studies have provided evidence that the mitochondrial proteins are synthesized by free polysomes in most cases and that the mitochondrial proteins completed on free cytoplasmic polysomes are released into the cytosolic compartment followed by rapid translocation into the mitochondria. It should be noted, however, that adrenodoxin was synthesized in vitro by both free and bound polysomes as a putative large precursor (41).

Many mitochondrial proteins were found to be synthesized as a precursor with an extra sequence, suggesting that the extra sequence may have a role possibly as a signal in the translocation of protein into the mitochondria. The mitochondrial matrix proteins in higher animals for which larger sizes of precursors have been identified are: malate de-hydrogenase (rat liver), carbamoyl phosphate synthase (rat liver), orni-thine transcarbamoylase (rat liver), aspartate aminotransferase (rat liver), serine aminotransferase (rat liver), phosphoenolpyruvate carboxykinase (chicken liver), ALA synthase (chick embryo liver and rat liver), and adrenodoxin (bovine adrenal cortext). However, some proteins are synthesized as polypeptides with no additional sequence. Proteolytic enzyme may constitute only a part of the protein-transporting machinery of the mitochondria. It is especially noteworthy that fumarase (32) and glycerate kinase (33, 34) in rat liver and phosphoenolpyruvate carboxy-kinase in bullfrog (35, 36) as well, occurring in the liver mitochondrial fraction were indistinguishable, if not absolutely identical, from their respective counterparts in the liver cytosol fraction by various analytical methods so far employed. At any rate, it seems conceivable to assume that the polypeptides to be incorporated into mitochondria have an addressing signal in a specific sequence of the peptides. It is interesting in this connection to note that Matsuura et al. (87) have recently shown that the precursor of liver cytochrome c differs from the mature cytochrome c only in that the precursor contains an $NH_2$-terminal methionine. The precursor of cytochrome c synthesized in vitro is incorporated into rat liver mitochondria and this incorporation is competed for by horse heart apocytochrome c, but not by the holocytochrome c. They further demonstrated that only one CNBr fragment of horse apocytochrome c, extending from residue 66 to the COOH-end of the molecule, could compete with the precursor for the transfer into mitochondria. An addressing signal may be contained in a specific segment of the cytochrome polypeptide. Inversely, mitochondria may have receptors somewhat specific to individual proteins to be transported into mitochondria (3, 87, 88), but so far no conclusive evidence is available for the occurrence of specific receptors on the mitochondrial surface.

There are several reports suggesting that energy may be required for the translocation of proteins through mitochondrial membranes (2, 24, 82, 88–90). This idea is based on the observations that the processing of precursors to the mitochondrial proteins was more or less prevented when the ATP level in the matrix of mitochondria was reduced by employing various types of uncouplers or inhibitors of oxidative phosphorylation. Although this idea is attractive, further studies appear to be needed to verify conclusively that protein translocation is really an energy-linked process.

The pool size of the cytosol precursor of the mitochondrial protein is supposed to be determined by both the rate of protein synthesis and the rate of its translocation into mitochondria. In hepatic porphyria anim-mals, the liver cytosol had a large pool of the precursor of ALA synthase (*42, 45*) and this cytosolic precursor was shown to be transferred into mitochondria with a half-disappearance time of about 20 min (*42, 48*). In addition, the intracellular translocation of hepatic ALA synthase from the cytosol into mitochondria appeared to be controlled by the heme concentration in the liver cell (*43, 45, 49, 62*). This is probably the only case in which the regulation of the intracellular translocation of mitochondrial matrix enzyme has been clearly demonstrated. On the other hand, the precursors of carbamoyl phosphate synthetase (*14, 16*) and ornithine transcarbamoylase (*16*) were reported to pass into mitochondria with a half-time of 1–2 min. It is interesting to note that the rate of intracellular translocation differs considerably for individual proteins.

## SUMMARY

Soluble matrix enzymes of the mitochondria are coded for by the nuclear gene and are synthesized outside mitochondria on cytoribosomes and then are transferred to the site of function across both the outer and inner mitochondrial membranes. Recent studies in various laboratories have shown that most of the mitochondrial matrix enzymes in higher animals are synthesized on membrane-free polysomes as precursors which are larger, though not always, than the mature form of the enzymes in the mitochondria. The precursors are transferred into mitochondria by a post-translational mechanism associated with proteolytic processing of the proteins to make the mature form of the enzymes and probably by an energy-dependent process. For some enzymes, however, the individual proteins distributing in the cytosol and mitochondria are indistinguishable each other by any criteria so far employed. Among those matrix enzymes in animals, hepatic δ-aminolevulinate (ALA) synthase is unique especially in the fact that the transfer of the enzyme from the cytosol into the mitochondria is regulated by the heme-mediated feedback mechanism. Synthesis of ALA synthase is also inhibited by heme possibly at both the transcriptional and post-transcriptional levels.

*Acknowledgment*
The experimental work reported from the authors' laboratory was supported in part by grants from the Ministry of Education, Science and

154 G. KIKUCHI AND N. HAYASHI

Culture, Japan, the Ministry of Health and Welfare, Japan, and the Yamanouchi Foundation for Metabolic Studies, Japan.

REFERENCES

1. Schatz, G. & Mason, T.L. (1974) *Annu. Rev. Biochem.* **43**, 51–87
2. Schatz, G. (1979) *FEBS Lett.* **103**, 203–211
3. Neupert, W. & Schatz, G. (1981) *Trends Biochem. Sci.* **6**, 1–4
4. Ades, I.Z. (1982) *Mol. Cell. Biochem.* **43**, 113–127
5. Harmey, M.A. & Neupert, W. (1979) *FEBS Lett.* **108**, 385–389
6. Godinot, C. & Lardy, H.A. (1973) *Biochemistry* **12**, 2051–2060
7. Kawajiri, K., Harano, T., & Omura, T. (1977) *J. Biochem.* **82**, 1403–1416
8. Kawajiri, K., Harano, T., & Omura, T. (1977) *J. Biochem.* **82**, 1417–1423
9. Bigham, R.W. & Campbell, P.N. (1972) *Biochem. J.* **126**, 211–215
10. Aziz, L.E., Chien, S.-M., Patel, H.V., & Freeman, K.B. (1981) *FEBS Lett.* **133**, 127–130
11. Dölken, G., Brdiczka, D., & Pette, D. (1973) *FEBS Lett.* **35**, 247–249
12. Shore, G.C., Carignan, P., & Raymond, Y. (1979) *J. Biol. Chem.* **254**, 3141–3144
13. Raymond, Y. & Shore, G.C. (1979) *J. Biol. Chem.* **254**, 9335–9338
14. Raymond, Y. & Shore, G.C. (1981) *J. Biol. Chem.* **256**, 2087–2090
15. Mori, M., Miura, S., Tatibana, M., & Cohen, P.P. (1979) *Proc. Natl. Acad. Sci. U.S.* **76**, 5071–5075
16. Mori, M., Morita, T., Ikeda, F., Amaya, Y., Tatibana, M., & Cohen, P.P. (1981) *Proc. Natl. Acad. Sci. U.S.* **78**, 6056–6060
17. Miura, S., Mori, M., Amaya, Y., Tatibana, M., & Cohen, P.P. (1981) *Biochem. Int.* **2**, 305–312
18. Mori, M., Morris, S.M., Jr., & Cohen, P.P. (1979) *Proc. Natl. Acad. Sci. U.S.* **76**, 3179–3183
19. Conboy, J.G., Kalousek, F., & Rosenberg, L.E. (1979) *Proc. Natl. Acad. Sci. U.S.* **76**, 5724–5727
20. Conboy, J.G. & Rosenberg, L.E. (1981) *Proc. Natl. Acad. Sci. U.S.* **78**, 3073–3077
21. Kraus, J.P., Conboy, J.G., & Rosenberg, L.E. (1981) *J. Biol. Chem.* **256**, 10739–10742
22. Mori, M., Miura, S., Tatibana, M., & Cohen, P.P. (1980) *J. Biochem.* **88**, 1829–1836
23. Mori, M., Miura, S., Tatibana, M., & Cohen, P.P. (1980) *Proc. Natl. Acad. Sci. U.S.* **77**, 7044–7048
24. Mori, M., Morita, T., Miura, S., & Tatibana, M. (1981) *J. Biol. Chem.* **256**, 8263–8266
25. Morita, T., Mori, M., Tatibana, M., & Cohen, P.P. (1981) *Biochem. Biophys. Res. Commun.* **99**, 623–629
26. Sakakibara, R., Huynh, Q.K., Nishida, Y., Watanabe, T., & Wada, H. (1980) *Biochem. Biophys. Res. Commun.* **95**, 1781–1788
27. Sakakibara, R., Kamisaki, Y., & Wada, H. (1981) *Biochem. Biophys. Res. Com-*

*mun.* 102, 235–242
28. Sonderegger, P., Jaussi, R., & Christen, P. (1980) *Biochem. Biophys. Res. Commun.* 94, 1256–1260
29. Oda, T., Ichiyama, A., Miura, S., Mori, M., & Tatibana, M. (1981) *Biochem. Biophys. Res. Commun.* 102, 568–573
30. Oda, T., Yanagisawa, M., & Ichiyama, A. (1982) *J. Biochem.* 91, 219–232
31. Hayashi, H., Katunuma, N., Chiku, K., Endo, Y., & Natori, Y. (1981) *J. Biochem.* 90, 1229–1232
32. Kobayashi, K., Yamanishi, T., & Tuboi, S. (1981) *J. Biochem.* 89, 1923–1931
33. Kitagawa, Y., Katayama, H., & Sugimoto, E. (1979) *Biochim. Biophys. Acta* 582, 260–275
34. Katayama, H., Kitagawa, Y., & Sugimoto, E. (1980) *J. Biochem.* 88, 765–773
35. Goto, Y., Shimizu, J., Okazaki, T., & Shukuya, R. (1979) *J. Biochem.* 86, 71–78
36. Goto, Y., Shimizu, J., & Shukuya, R. (1980) *J. Biochem.* 88, 1239–1249
37. Goto, Y., Ohki, Y., Shimizu, J., & Shukuya, R. (1981) *Biochim. Biophys. Acta* 657, 383–389
38. Ohki, Y., Goto, Y., & Shukuya, R. (1981) *Biochim. Biophys. Acta* 661, 230–234
39. Kochi, H. & Kikuchi, G. (1979) *Seikagaku* (in Japanese) 51, 894
40. Kochi, H., Serizawa, K., & Kikuchi, G. (1980) *J. Biochem.* 88, 895–904
41. Nabi, N. & Omura, T. (1980) *Biochem. Biophys. Res. Commun.* 97, 680–686
42. Hayashi, N., Yoda, B., & Kikuchi, G. (1969) *Arch. Biochem. Biophys.* 131, 83–91
43. Hayashi, N., Kurashima, Y., & Kikuchi, G. (1972) *Arch. Biochem. Biophys.* 148, 10–21
44. Nakakuki, M., Yamauchi, K., Hayashi, N., & Kikuchi, G. (1980) *J. Biol. Chem.* 255, 1738–1745
45. Yamauchi, K., Hayashi, N., & Kikuchi, G. (1980) *J. Biol. Chem.* 255, 1746–1751
46. Yamauchi, K., Hayashi, N., & Kikuchi, G. (1980) *FEBS Lett.* 115, 15–18
47. Hayashi, N., Terasawa, M., & Kikuchi, G. (1980) *J. Biochem.* 88, 921–926
48. Hayashi, N., Terasawa, M., Yamauchi, K., & Kikuchi, G. (1980) *J. Biochem.* 88, 1537–1543
49. Yamamoto, M., Hayashi, N., & Kikuchi, G. (1981) *Arch. Biochem. Biophys.* 209, 451–459
50. Ohashi, A. & Kikuchi, G. (1972) *Arch. Biochem. Biophys.* 153, 34–46
51. Whiting, M.J. (1976) *Biochem. J.* 158, 391–400
52. Brooker, J.D., May, B.K., & Elliott, W.H. (1980) *Eur. J. Biochem.* 106, 17–24
53. Ades, I.Z. & Harpe, K.G. (1981) *J. Biol. Chem.* 256, 9329–9333
54. Granick, S. & Sassa, S. (1971) in *Metabolic Pathways* (Vogel, H.J., ed.) Vol. 5, pp. 77–141, Academic Press, New York
55. Granick, S. & Beale, S.I. (1978) *Adv. Enzymol.* 46, 33–203
56. Beattie, D.S. & Stuchell, R.N. (1970) *Arch. Biochem. Biophys.* 139, 291–297
57. Gross, S.R. & Hutton, J.J. (1971) *J. Biol. Chem.* 246, 606–614
58. Scholnick, P.L., Hammaker, L.E., & Marver, H.S. (1972) *J. Biol. Chem.* 247, 4126–4131
59. Whiting, M.J. & Elliott, W.H. (1972) *J. Biol. Chem.* 247, 6818–6826
60. Igarashi, J., Hayashi, N., & Kikuchi, G. (1976) *J. Biochem.* 80, 1091–1099

61. Tomita, Y., Ohashi, A., & Kikuchi, G. (1974) *J. Biochem.* **75**, 1007–1015
62. Kikuchi, G. & Hayashi, N. (1981) *Mol. Cell. Biochem.* **37**, 27–41
63. Yamamoto, M., Hayashi, N., & Kikuchi, G. (1982) *Biochem. Biophys. Res. Commun.* **105**, 985–990
64. Kurashima, Y., Hayashi, N., & Kikuchi, G. (1970) *J. Biochem.* **66**, 863–865
65. Marver, H.S., Collins, A., Tschudy, D.P., & Rechcigl, M., Jr. (1966) *J. Biol. Chem.* **241**, 4323–4329
66. Tschudy, D.P., Marver, H.S., & Collins, A. (1965) *Biochem. Biophys. Res. Commun.* **21**, 480–487
67. Meyer, U.A., Strand, L.J., Doss, M., Rees, A.C., & Marver, H.S. (1972) *New Engl. J. Med.* **286**, 1277–1282
68. Sassa, S. & Granick, S. (1970) *Proc. Natl. Acad. Sci. U.S.* **67**, 517–522
69. Strand, L.J., Manning, J., & Marver, H.S. (1972) *J. Biol. Chem.* **247**, 2820–2827
70. Tyrrell, D.L.J. & Marks, G.S. (1972) *Biochem. Pharmacol.* **21**, 2077–2093
71. Srivastava, G., Brooker, J.D., May, B.K., & Elliott, W.H. (1980) *Biochem. J.* **188**, 781–788
72. Badawy, A.A.-B. & Evans, M. (1973) *Biochem. J.* **136**, 885–892
73. Badawy, A.A.-B. & Evans, M. (1975) *Biochem. J.* **150**, 511–520
74. Chua, N.-H. & Schmidt, G.W. (1979) *J. Cell Biol.* **81**, 461–483
75. Mihara, K., Omura, T., Harano, T., Brenner, S., Fleischer, S., Rajagopalan, K.V., & Blobel, G. (1982) *J. Biol. Chem.* **257**, 3355–3358
76. Gaitskhoki, V.S., Kisselev, O.I., Klimov, N.A., Monakhov, N.K., Mukha, G.V., Schwartzman, A.L., & Neifakh, S.A. (1974) *FEBS Lett.* **43**, 151–154
77. Shore, G.C. & Tata, J.R. (1977) *J. Cell Biol.* **72**, 726–743
78. Kellems, R.E., Allison, V.F., & Butow, R.A. (1974) *J. Biol. Chem.* **249**, 3297–3303
79. Ades, I.Z. & Butow, R.A. (1980) *J. Biol. Chem.* **255**, 9918–9924
80. Ades, I.Z. & Butow, R.A. (1980) *J. Biol. Chem.* **255**, 9925–9935
81. Miura, S., Mori, M., Amaya, Y., & Tatibana, M. (1982) *Eur. J. Biochem.* **122**, 641–647
82. Morita, T., Miura, S., Mori, M., & Tatibana, M. (1982) *Eur. J. Biochem.* **122**, 501–509
83. Huynh, Q.K., Sakakibara, R., Watanabe, T., & Wada, H. (1980) *J. Biochem.* **88**, 231–239
84. Marra, E., Doonan, S., Saccone, C., & Quagliariello, E. (1977) *Biochem. J.* **164**, 685–691
85. Marra, E., Doonan, S., Saccone, C., & Quagliariello, E. (1978) *Eur. J. Biochem.* **83**, 427–435
86. Marra, E., Passarella, S., Doonan, S., Quagliariello, E., & Saccone, C. (1980) *FEBS Lett.* **122**, 33–36
87. Matsuura, S., Arpin, M., Hannum, C., Margoliash, E., Sabatini, D.D., Morimoto, T. (1981) *Proc. Natl. Acad. Sci. U.S.* **78**, 4368–4372
88. Zimmermann, R., Hennig, B., & Neupert, W. (1981) *Eur. J. Biochem.* **116**, 455–460
89. Nelson, N. & Schatz, G. (1979) *Proc. Natl. Acad. Sci. U.S.* **76**, 4365–4369
90. Zimmermann, R. & Neupert, W. (1980) *Eur. J. Biochem.* **109**, 217–229

# Biogenesis of the Membrane
# of Endoplasmic Reticulum

TSUNEO OMURA

*Department of Biology, Faculty of Science, Kyushu University, Fukuoka 812, Japan*

Biomembranes are dynamic structures whose lipid and protein constituents are continuously synthesized and degraded in living cells. Various types of membranes exist together in eukaryotic cells, and the central problem of membrane biogenesis is the mechanism by which characteristic enzyme compositions of those different membranes are maintained in the face of rapid turnover of the membrane constituents. Another important factor to be elucidated is the mechanism which effects clear asymmetric distribution of various proteins in the membranes.

The membrane of endoplasmic reticulum, which is transformed into microsomal vesicles upon the homogenization of cells, is particularly abundant in the cells of certain animal tissues including liver. Since many integral membrane proteins of endoplasmic reticulum have already been purified and well characterized, this membrane system is a suitable research subject to use in studying those basic problems of membrane biogenesis in animal cells. The sidedness of various membrane proteins in microsomal vesicles has also been studied in recent years.

Since the first pioneering study by Dallner *et al.* (*1, 2*) on the formation of endoplasmic reticulum in the hepatocytes of newborn rats, the biogenesis of this membrane system has been studied mostly with mammalian liver cells, in which both smooth-surfaced and rough-surfaced endoplasmic reticulum are abundant. Although we are still far from a clear understanding of the mechanism of formation and degradation of endoplasmic reticulum membrane, much new information has been accumulated in the past decade. This short review summarizes the available information on the biosynthesis, integration into membrane, and turnover of microsomal proteins in the liver cell.

## I. TURNOVER OF MICROSOMAL MEMBRANE PROTEINS IN ANIMAL LIVER

The dynamic nature of endoplasmic reticulum is most clearly shown by the rapid turnover of its membrane constituents. The average half-life of microsomal membrane proteins in rat liver is 50–60 h ($3$–$5$), and that of membrane phospholipids is 30–40 h ($3$). Apparently, the synthesis and the degradation of phospholipids in microsomal membrane are not tightly coupled with those of the bulk protein constituents of the membrane. It was also found that the phospholipids of microsomal membrane are in a dynamic state of reversible transfer with those of other membranes in the liver cell owing to the presence of phospholipid exchange proteins in the cytosol ($6$).

A noteworthy feature of the turnover of microsomal enzymes is their independent random turnover. Different turnover rates of microsomal

TABLE I.  Half-lives of microsomal enzymes in rat liver.

| Enzymes | Half-lives | References |
|---|---|---|
| Stearoyl CoA desaturase | 3–4 h | ($8,9$) |
| β-Hydroxymethylglutaryl CoA reductase | 4 // | ($10$) |
| Cytochrome P-450 (cholesterol 7α-hydroxylase) | 4 // | ($11$) |
| Squalene synthetase | 12 // | ($12$) |
| Glutathione-insulin transhydrogenase | 26 // | ($13$) |
| Nucleoside diphosphatase | 30 // | ($14$) |
| Cytochrome P-450 (phenobarbital-inducible form) | 40 // | ($15$) |
| NADPH-cytochrome $c$ reductase | 70 // | ($4$) |
| Aryl acylamidase | 100 // | ($16$) |
| Cytochrome $b_5$ | 100 // | ($4$) |
| NADH-cytochrome $b_5$ reductase | 6 days | ($17$) |
| Protein disulfide isomerase | 7 // | ($18$) |
| NAD glycohydrolase | 18 // | ($19$) |

enzymes were first noticed for cytochrome $b_5$ and NADPH-cytochrome $c$ reductase (3), and then confirmed for other protein components of the membrane (5, 7). Table I summarizes reported half-lives of microsomal enzymes in rat liver.

The turnover rates of microsomal enzymes were usually determined by labeling the enzymes *in vivo* with radioactive amino acids and then measuring the decay of their radioactivities. The half-lives of some enzymes having high turnover rates were determined from the decrease of enzyme activities after cessation of enzyme synthesis. Although the former method is subject to errors caused by reutilization of labeled amino acids *in vivo*, giving longer half-lives than actual values (5, 20–22), and the latter method could be disturbed by activation and inactivation of enzymes, observed wide differences in the half-lives of microsomal enzymes, ranging from a few hours to several days as shown in Table I, clearly indicate independent random turnover of various protein constituents of microsomal membrane. The membrane of endoplasmic reticulum is apparently a loose dynamic structure, and each of its protein constituents can be incorporated into or removed from the membrane without affecting others.

We have only limited information on the mechanism of turnover of microsomal enzymes in the liver cell. Although the role of lysosomes in the degradation of cellular proteins is well recognized (23, 24) and morphological observations often confirmed the engulfment of endoplasmic reticulum membrane by lysosomes in liver cells (23, 25, 26), the heterogeneous turnover rates of microsomal enzymes do not support the major contribution of the autophagic process to the degradation of endoplasmic reticulum in the liver cells of normal animals. If we assume that various microsomal enzymes are uniformly distributed in the membrane of endoplasmic reticulum, the autophagic digestion of the membrane as a whole will result in the uniform turnover rate of various microsomal enzymes. Subfractionation of rat liver microsomes by differential or isopicnic centrifugations gave evidence for non-uniform distribution of enzymes among microsomal vesicles (27–29), suggesting some extent of heterogeneity of the membranes of both rough-surfaced and smooth-surfaced endoplasmic reticulum. A similar result was also obtained by the fractionation of rat liver microsomes with an immunoadsorption method (30). However, we find some pairs of microsomal enzymes whose components are intimately associated with each other in the catalysis of electron transfer reactions and yet have widely different turnover rates, *e.g.*, stearoyl CoA desaturase and cytochrome $b_5$, NADPH-cytochrome $c$ reductase and cytochrome P-450 (Table I). If the autophagic process is mainly responsible for the degradation of the membrane of endoplasmic reticulum, it is difficult to explain the independent turnover of the components of such enzyme pairs.

Autophagic process seems to be highly activated under certain physiological conditions (23), however, and may contribute significantly to the degradation of intracellular membranes in the liver cell. When rats were treated with phenobarbital for several days and then the treatment was interrupted, a remarkable increase of autophagic vacuoles containing smooth membranes was observed in the liver cells of the treated animals (31). The increase of autophagic vacuoles coincided with the regression of smooth endoplasmic reticulum proliferated during the drug administration, and strongly suggested significant contribution of the autophagic process to the degradation of microsomal membrane proteins in the liver of the treated animals (31). It is likely that the autophagic digestion of endoplasmic reticulum also contributes to the turnover of microsomal membrane proteins (25, 26), but its contribution varies significantly according to the physiological conditions of the animals.

In order to explain the observed random turnover rates of microsomal membrane proteins, we can assume the participation of non-lysosomal protease(s) in the membrane of microsomes or in the cytosol, which attacks the membrane-bound enzymes and determines their turnover rates. Since good correlations between the *in vitro* susceptibilities to proteases and the *in vivo* turnover rates are reported for various soluble enzymes (32, 33), it is tempting to assume the role of such extra-lysosomal proteases in the degradation of microsomal enzymes. As was first shown for cytochrome $b_5$ (34), microsomal membrane proteins are generally amphipatic, and the protein molecules are bound to the membrane by burying their hydrophobic portions in the membrane, whereas their hydrophilic portions stick out of the membrane into the surrounding aqueous environment. Some microsomal enzymes, including cytochrome $b_5$ (35) and NADH-cytochrome $b_5$ reductase (37, 38), consist of functionally and conformationally distinct domains, and their hydrophilic domains can be easily dissected from the membrane-binding hydrophobic domains by proteases. Moreover, some microsomal enzymes show significant specificity to proteases in the proteolytic solubilization from the membrane. NADH-cytochrome $b_5$ reductase can be efficiently solubilized from microsomes by lysosomal cathepsins but is relatively resistant to solubilization by trypsin, whereas NADPH-cytochrome $c$ reductase and cytochrome $b_5$ are readily solubilized by trypsin but not by lysosomal digestion (39).

The presence of soluble NADH-cytochrome $b_5$ reductase and cytochrome $b_5$ in erythrocytes (40, 41) whose properties are almost the same as their counterparts solubilized from liver microsomes by protease treatment (42, 43), suggests proteolytic solubilization of these membrane proteins from disappearing endoplasmic reticulum during erythroid maturation. These observations with erythrocytes seem to indicate the detachment of some membrane proteins from microsomes by limited proteolysis in the process of their degradation in situ. However, no clear evidence for the involvement of such a proteolytic solubilization step in the normal turnover of microsomal enzymes in the liver cell has yet been reported.

When an enzyme is found in the membranes of different cell organelles, studies on its turnover could give us important information concerning the mechanism of degradation of membrane proteins. Some microsomal enzymes are distributed on outer mitochondrial membrane, nuclear envelope, Golgi apparatus, etc. in the liver cell, and seem to be suited for studies to examine whether or not the same enzyme bound to different types of membrane shows an identical turnover rate. NADH-cytochrome $b_5$ reductase is one of such microsomal enzymes showing multimodal intracellular distribution, and the molecular identity of the reductases found in different cell organelles of the liver cell is confirmed (44, 45). Okada and Omura (46) reported an identical half-life for microsomal and mitochondrial NADH-cytochrome $b_5$ reductases in rat liver, and suggested that the turnover rate of this enzyme is not affected by the properties of the membrane with which it binds. On the other hand, Borgese et al. (47) reported that the reductases of microsomal membrane and Golgi apparatus showed faster turnover rates than their counterpart in outer mitochondrial membrane. More studies on this and other microsomal enzymes are needed.

## II.  BIOSYNTHESIS OF MICROSOMAL MEMBRANE PROTEINS

When a new enzyme appears in the membrane of endoplasmic reticulum during cell development or enzyme induction, it usually appears first in the rough-surfaced portion and then in the smooth-surfaced portion. This was first shown by Dallner *et al.* (*2*) for the appearance of glucose-6-phosphatase and NADPH-cytochrome *c* reductase in developing rat hepatocytes. Histochemical staining of glucose-6-phosphatase activity in the cells of fetal rat livers (*48*) also confirmed its initial appearance in rough-surfaced endoplasmic reticulum. In the induced increase of cytochrome P-450 and NADPH-cytochrome *c* reductase in the livers of phenobarbital-treated rats, the enzymes of rough microsomes increased first followed by those of smooth microsomes (*49*). Stearoyl CoA desaturation activity of rat liver microsomes, which can be remarkably induced by refeeding fasted animals, also showed a much faster increase in rough microsomes than in smooth microsomes during the induction (*9*).

The incorporation of newly synthesized microsomal enzymes into membrane at the rough-surfaced region of endoplasmic reticulum was further confirmed for cytochrome $b_5$ (*50, 51*), NADPH-cytochrome *c* reductase (*51*), and cytochrome P-450 (*52*) by label-chase experiments using whole animals. The initial rates of incorporation of injected radioactive amino acids were significantly faster for the enzymes of rough microsomes than of smooth microsomes, and then radioactive enzyme molecules gradually distributed evenly between the two types, suggesting reversible migration of newly synthesized enzyme molecules from the site of incorporation to other parts of the continuous endoplasmic reticulum membrane. In the case of NADH-cytochrome $b_5$ reductase, however, one paper (*46*) reported a faster incorporation of radioactivity into the enzyme of rough microsomes, whereas another paper (*47*) reported an identical rate of incorporation of a labeled amino acid into the enzymes of both microsomal subfractions.

Since these microsomal enzymes are hydrophobic proteins having highly hydrophobic portions in the molecules, they are not freely soluble in aqueous media and have a strong tendency to bind to membranes (*53–55*). The appearance of newly synthesized enzyme molecules in the rough-surfaced portion of endoplasmic reticulum *in vivo* can be explained by their preferential synthesis by the membrane-bound ribosomes in the rough-surfaced portion of endoplasmic reticulum in the liver cell.

Studies on the synthesis of microsomal enzymes in the liver cell, where both free and membrane-bound ribosomes are abundant, have therefore been focused on the contributions of these two types of ribosomes to the synthesis of specific enzyme proteins whose molecular prop-

TABLE II.  Contributions of free and membrane-bound ribosomes to the synthesis of microsomal enzymes in the liver cell.

| Enzymes | Synthesized by | References |
|---|---|---|
| NADH-cytochrome $b_5$ reductase | Free Rs | (56,57) |
| NADPH-cytochrome $c$ reductase | Free Rs | (58) |
| // | Free & bound Rs | (59) |
| // | Bound Rs | (57,60) |
| Cytochrome $b_5$ | Bound Rs | (50) |
| // | Loosely-bound Rs | (61) |
| // | Free & loosely-bound Rs | (60,62) |
| // | Free Rs | (57,63) |
| Cytochrome P-450 | Bound Rs | (52,64) |
| Epoxide hydratase | Bound Rs | (57,65) |
| Disulfide interchange enzyme | Bound Rs | (18) |

erties were known. As the components of the microsomal electron transfer system, cytochrome P-450, cytochrome $b_5$, NADH-cytochrome $b_5$ reductase, and NADPH-cytochrome $c$ reductase, have been intensively studied from enzymological points of view and well characterized, they were the primary choice of many biosynthetic studies. Table II summarizes the results reported in the literature concerning the contributions of free and membrane-bound ribosomes to the synthesis of microsomal enzymes in the liver cell.

Although we notice some discrepancies among the reported data, it is apparent that not all microsomal enzymes are synthesized by membrane-bound ribosomes which are tightly associated with the membrane of rough-surfaced endoplasmic reticulum. These discrepancies could be due to differences in the experimental conditions employed: feeding conditions of animals, salt concentrations of the sucrose media used in the isolation of free ribosomes and rough microsomes from liver homogenates, methods of detection of the synthesis of specific proteins, *etc.* Starved rats were used in some studies, whereas animals were not fasted until killed in other studies. Starvation of animals alters the populations of free and bound ribosomes in the liver cell. Salt concentrations used in the sucrose media ranged from 25 mM KCl (66) to as high as 250 mM KCl (67). In the presence of high concentrations of KCl, some portions of the bound ribosomes detach from the membrane of rough microsomes (68), and these are usually called loosely-bound ribosomes. Earlier studies employed the detection of the ribosome-associated nascent peptides of specific enzymes by immunochemical methods, which was expected to give direct estimation of the populations of free and membrane-bound ribosomes participating in the synthesis of the enzymes *in situ*. More recent studies, on the other hand, utilize *in vitro* translation of isolated polysomes or mRNA from the polysomes to detect the completed peptides of specific enzymes in the translation products.

Studies with other types of animal cells also indicated lack of absolute segregation between free and membrane-bound ribosomes in the synthesis of microsomal membrane proteins. Some microsomal enzymes (heme oxygenase in cultured alveolar macrophages (*69*), *etc.*) are synthesized by free ribosomes, whereas some others (calcium-transport ATPase in the muscle cells of chick embryos (*70*) and neonatal rats (*71*), *etc.*) are apparently synthesized by membrane-bound ribosomes. We may conclude that free ribosomes also participate in the synthesis of some microsomal membrane proteins, although the contribution of membrane-bound ribosomes to the synthesis of bulk microsomal membrane proteins seems to be far greater than that of free ribosomes (*72*).

In contrast with secretory proteins, which are synthesized as larger precursor peptides (*73, 74*) by the membrane-bound ribosomes of rough-surfaced endoplasmic reticulum to be segregated into the lumen (*75, 76*), microsomal enzymes are generally synthesized as mature-size peptides. *In vitro* translation of polysomes or mRNA confirmed the synthesis of cytochrome $b_5$ (*57, 60, 62, 63, 77*), NADH-cytochrome $b_5$ reductase (*57*), NADPH-cytochrome $c$ reductase (*57, 60*), epoxide hydratase (*57, 65, 78*), and certain forms of cytochrome P-450 (*64, 79, 80*) as peptides which were indistinguishable in size from corresponding mature enzymes. A form of microsomal cytochrome P-450 in the liver of 3-methylcholan-threne-treated rats was reported to be synthesized as a large precursor form (*79*), but this observation has not yet been confirmed by other investigators.

## III.  INTEGRATION OF NEWLY SYNTHESIZED PROTEINS INTO ENDOPLASMIC RETICULUM MEMBRANE

Independent turnover of various membrane proteins of endoplasmic reticulum in the liver cell indicates random incorporation of newly synthesized proteins into the existing membrane. Since the majority of microsomal membrane proteins seem to be synthesized by membrane-bound

TABLE III. Sidedness of various membrane-bound enzymes in microsomal vesicles.

| Enzymes | Sidedness in microsomes | References |
|---|---|---|
| Acyl CoA-cholesterol acyltransferase | Outside | (81) |
| Acyl CoA lygase | " | (82) |
| Acyl transferase | " | (82) |
| Choline phosphotransferase | " | (82) |
| Cytochrome $b_5$ | " | (3,83) |
| Cytochrome P-450 | " | (83,84) |
| Ethanolamine phosphotransferase | " | (82) |
| Glutathione S-transferase | " | (86) |
| $\beta$-Hydroxymethylglutaryl CoA reductase | " | (87) |
| NADH-cytochrome $b_5$ reductase | " | (44,85) |
| NADPH-cytochrome $c$ reductase | " | (3,83) |
| Stearoyl CoA desaturase | " | (88) |
| Arylsulfatase C | Trans-membrane | (89) |
| Ribophorin | " | (90) |
| Aryl acylamidase | Inside | (91) |
| Glucose-6-phosphatase | " | (92,93) |
| Lysyl hydroxylase | " | (94) |
| Nucleoside diphosphatase | " | (14) |

ribosomes and incorporated immediately into the rough-surfaced portion of endoplasmic reticulum where the ribosomes are associated, simple hydrophobic attraction between newly synthesized peptides and the membrane can account for the observed efficient incorporation of various hydrophobic proteins into the cytoplasmic surface of endoplasmic reticulum membrane. However, we find some microsomal enzymes located on the luminal surface of isolated microsomal vesicles (28). Trans-membranous disposition of the protein molecules has also been confirmed with a few microsomal enzymes. Table III summarizes available information about the sidedness of various membrane-bound enzymes in microsomal vesicles.

Although conflicting reports have also been published concerning the sidedness of some enzymes in microsomal vesicles (cytochrome P-450 and NADPH-cytochrome $c$ reductase (96), nucleoside diphosphatase (95), etc.), clear segregation of various microsomal enzymes to the cytoplasmic or luminal side of the membrane of endoplasmic reticulum as shown in Table III is now generally accepted (28). Such an asymmetric distribution of membrane proteins in endoplasmic reticulum is effected at the step of incorporation of newly synthesized proteins into the membrane. Various phospholipids are also asymmetrically distributed on cytoplasmic and luminal sides of the membrane (28, 97, 98).

Penetration of growing nascent peptides across the membrane was first confirmed for secretory proteins (75, 76), which are synthesized by the membrane-bound ribosomes of rough-surfaced endoplasmic reticulum and discharged into the lumen. This remarkable property of the nascent peptides of secretory proteins is due to the presence of particular short amino acid sequences, which are called "signal peptides," at their amino terminal portions (73, 74). Signal peptide sequences of a number of secretory proteins have already been determined, and a common feature of signal peptides is an uninterrupted stretch of uncharged, mostly hydrophobic, amino acids preceded by a highly charged amino acid, lysine or arginine in most cases, near the amino terminal end (99). The signal peptides are split by a protease at the luminal surface of the membrane of endoplasmic reticulum, releasing secretory proteins into the lumen (73). Type I in Fig. 1 shows this process according to a recent scheme (99). If the signal sequence is not split by the protease, the completed protein molecule will remain attached to the luminal surface of the membrane to become an inside-located membrane protein (type II in Fig. 1).

Although available information about the conformation of microsomal enzymes in the membrane is limited, Fig. 1 shows schematically how the peptides of inside-located (type II), trans-membranous (types III

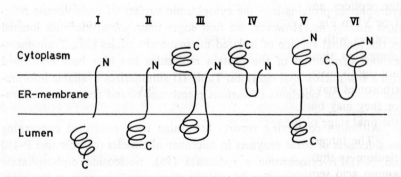

Fig. 1. Various types of integral membrane proteins of endoplasmic reticulum. Penetrations of peptides across the membrane of endoplasmic reticulum are schematically shown. N and C denotes amino terminal end and carboxy terminal end of a peptide, respectively.

and V), and outside-located (types IV and VI) microsomal enzymes are disposed in the membrane. A thermodynamic consideration (*100*) explains the essential importance of hydrophobic signal sequences in the penetration of peptides across membranes, and the same seems to be applicable to membrane-bound proteins (*101*).

Types II and IV in Fig. 1 show membrane-bound proteins each having a hydrophobic sequence at the amino terminal portion of the peptide. Several forms of microsomal cytochrome P-450 have been shown to have highly hydrophobic sequences at the amino terminal portion (*102–104*), and they are possibly bound to the membrane as shown by type IV. Presence of a hydrophobic sequence at the amino terminal portion was also shown for NADPH-cytochrome $c$ reductase (*105*) and epoxide hydratase (*106*). However, a conflicting observation was reported concerning the membrane-binding portion of NADPH-cytochrome $c$ reductase (*107*).

Trans-membrane proteins must have hydrophobic segments in the middle of the peptides as shown by types III and V in Fig. 1. No sequence data is available for microsomal trans-membrane proteins, but band III protein of the erythrocyte membrane seems to have a disposition of the peptide as shown by type V when synthesized by membrane-bound ribosomes and incorporated into the membrane of rough-surfaced endoplasmic reticulum in erythroid cells (*108*).

Cytochrome $b_5$ (*109*) and NADH-cytochrome $b_5$ reductase (*110*) have membrane-binding hydrophobic segments at the carboxyl end of the peptides, and these are possibly bound to the membrane as shown by type VI in Fig. 1. In accordance with the notion that the binding of the peptides with the membrane should occur at the last stage of peptide elongation, these proteins seem to be synthesized by ribosomes which are not tightly associated with the membrane of endoplasmic reticulum. The ribosomes may be free in the cytoplasm when the peptides are completed, or they may bind loosely to the membrane of endoplasmic reticulum at the final stage of peptide elongation.

The integration of proteins into the membrane of endoplasmic reticulum is thus explained by the presence of particular hydrophobic amino acid sequences in the peptides. However, an important problem still remains unsolved, that is, how various microsomal proteins are selectively incorporated into the membrane of endoplasmic reticulum. *In vitro* transfer of some microsomal proteins, cytochrome $b_5$ (*111*) and

NADH-cytochrome $b_5$ reductase (112), between artificial phospholipid vesicles was once reported, but later observations (113) did not support an inter-membrane transfer of these enzymes between biological membranes. Therefore, the incorporation of newly synthesized proteins into membranes has decisive importance in determining the intracellular distribution of various integral membrane proteins among membranous cell organelles.

The same problem is also shared by secretory proteins. The translation of mRNA in the cytoplasm of animal cells seems to start on free ribosomes (114, 115), but the growing nascent peptides of secretory proteins recognize the membrane of endoplasmic reticulum and bind to it, resulting in efficient co-translational translocation of the peptides across the membrane into the lumen (73, 74). They also bind to the outer membrane of nuclei, but do not bind to other membranes including outer mitochondrial membrane. Recently, participation of a protein factor, called "signal recognizing protein," has been proposed for the recognition of the membrane of endoplasmic reticulum by the growing nascent peptides of secretory proteins (116, 117). Since hydrophobic amino acid sequences at the amino terminal portions of some microsomal proteins resemble those of secretory proteins (102), it is highly likely that a "signal recognizing protein" or similar protein component mediates the selective binding of the ribosomes carrying the growing nascent peptides of microsomal proteins to the membrane of endoplasmic reticulum. This interesting possibility has much relevance with the basic mechanism of membrane biogenesis in eukaryotic cells and is to be elucidated by future studies.

## SUMMARY

Based on the available information concerning the turnover and biosynthesis of microsomal membrane proteins in the liver cell, we can outline the mechanism of formation of the membrane of endoplasmic reticulum, which maintains its characteristic structure and composition in the face of rapid turnover of the constituents.

Although some microsomal proteins seem to be synthesized by ribosomes which are not tightly associated with the membrane, the majority of microsomal membrane proteins are synthesized by the membrane-bound ribosomes of rough-surfaced endoplasmic reticulum.

The growing nascent peptides bind with the membrane and remain bound during their elongation. When the enzyme molecules are completed, they thus appear in the membrane of the rough-surfaced portion of endoplasmic reticulum, and then migrate into other parts of the continuous membrane by diffusion to be equally distributed between the membranes of rough- and smooth-surfaced endoplasmic reticula. The asymmetric distribution of various microsomal enzymes between cytoplasmic and luminal sides of the membrane can be explained by the penetration of the nascent peptides of a particular class of such enzymes whose catalytic portions are present on the luminal side, across the membrane, as is true of the nascent peptides of secretory proteins. On the other hand, the nascent peptides of cytoplasmic side-located enzymes also attach to the membrane but do not pass through it. Elucidation of the complete amino acid sequences of various microsomal enzymes and the disposition of their peptides in the membrane will give us satisfactory explanation why the nascent peptides of some enzymes penetrate the membrane whereas those of others do not.

# REFERENCES

1. Dallner, G., Siekevitz, P., & Palade, G.E. (1966) *J. Cell Biol.* **30**, 73–96
2. Dallner, G., Siekevitz, P., & Palade, G.E. (1966) *J. Cell Biol.* **30**, 97–117
3. Omura, T., Siekevitz, P., & Palade, G.E. (1967) *J. Biol. Chem.* **242**, 2389–2396
4. Kuriyama, Y., Omura, T., Siekevitz, P., & Palade, G.E. (1969) *J. Biol. Chem.* **244**, 2017–2026
5. Arias, I.M., Doyle, D., & Schimke, R.T. (1969) *J. Biol. Chem.* **244**, 3303–3315
6. Wirtz, K.W.A. (1974) *Biochim. Biophys. Acta* **344**, 95–117
7. Dehlinger, P.J. & Schimke, R.T. (1971) *J. Biol. Chem.* **246**, 2574–2583
8. Gelhorn, A. & Benjamin, W. (1966) *Biochim. Biophys. Acta* **116**, 460–466
9. Oshino, N. & Sato, R. (1972) *Arch. Biochem. Biophys.* **149**, 369–377
10. Edwards, P.A. & Gould, R.G. (1972) *J. Biol. Chem.* **247**, 1520–1524
11. Gielen, J., Van Cantfort, J., Robaye, B., & Renson, J. (1975) *Eur. J. Biochem.* **55**, 41–48
12. Slakey, L.L., Craig, M.C., Beytia, E., Briedis, A., Feldruegge, D.H., Dugan, R.E., Qureshi, A.A., Subbarayan, C., & Porter, J.W. (1972) *J. Biol. Chem.* **247**, 3014–3022
13. Hern, E.P. & Valandani, P.T. (1980) *J. Biol. Chem.* **255**, 697–703
14. Kuriyama, Y. (1972) *J. Biol. Chem.* **247**, 2979–2988
15. Fujii-Kuriyama, Y., Mikawa, R., Tashiro, Y., Sakai, M., & Muramatsu, M. (1978) *Seikagaku* (in Japanese), **50**, 870

16. Akao, T. & Omura, T. (1972) *J. Biochem.* **72**, 1257–1259
17. Okada, Y. & Omura, T. (1978) *J. Biochem.* **83**, 1039–1048
18. Ohba, H., Harano, T., & Omura, T. (1981) *J. Biochem.* **89**, 901–907
19. Bock, K.W., Siekevitz, P., & Palade, G.E. (1971) *J. Biol. Chem.* **246**, 188–195
20. Swick, R.W. (1958) *J. Biol. Chem.* **231**, 751–764
21. Poole, B. (1971) *J. Biol. Chem.* **246**, 6587–6591
22. Swick, R.W. & Ip, M.M. (1974) *J. Biol. Chem.* **67**, 259–266
23. De Duve, C. & Wattiaux, R. (1966) *Annu. Rev. Physiol.* **28**, 435–492
24. Goldberg, A.L. & St. John, A.C. (1976) *Annu. Rev. Biochem.* **45**, 747–803
25. Marzella, L., Ahlberg, J., & Glaumann, H. (1981) *Virchows Arch. Cell Pathol.* **36**, 219–234
26. Furuno, K., Ishikawa, T., & Kato, K. (1982) *J. Biochem.* **91**, 1485–1494
27. Dallman, P.R., Dallner, G., Bergstrand, A., & Ernster, L. (1969) *J. Cell Biol.* **41**, 357–377
28. DePierre, J.W. & Dallner, G. (1975) *Biochem. Biophys. Acta* **415**, 411–472
29. Beaufay, H., Amar-Costesec, A., Feytmans, E., Thines-Sempoux, D., Wibo, M., Robbi, M., & Berthet, J. (1974) *J. Cell Biol.* **61**, 213–231
30. Kawajiri, K., Ito, A., & Omura, T. (1977) *J. Biochem.* **81**, 779–789
31. Bolender, R.P. & Weibel, E.R. (1973) *J. Cell Biol.* **56**, 746–761
32. Segal, H.L., Winkler, J.R., & Miyagi, M.P. (1974) *J. Biol. Chem.* **249**, 6364–6365
33. Goldberg, A.L. & Dice, J.F. (1974) *Annu. Rev. Biochem.* **43**, 835–869
34. Ito, A. & Sato, R. (1968) *J. Biol. Chem.* **243**, 4922–4923
35. Spatz, L. & Strittmatter, P. (1971) *Proc. Natl. Acad. Sci. U.S.* **68**, 1042–1046
36. Visser, L., Robinson, N.C., & Tanford, C. (1975) *Biochemistry* **14**, 1194–1199
37. Mihara, K. & Sato, R. (1972) *J. Biochem.* **71**, 725–735
38. Spatz, L. & Strittmatter, P. (1973) *J. Biol. Chem.* **248**, 793–799
39. Takesue, S. & Omura, T. (1970) *J. Biochem.* **67**, 259–266
40. Passon, P.G., Reed, D.W., & Hultquist, D.E. (1972) *Biochim. Biophys. Acta* **275**, 51–61
41. Passon, P.G. & Hultquist, D.E. (1972) *Biochim. Biophys. Acta* **275**, 62–73
42. Douglas, R.H. & Hultquist, D.E. (1978) *Proc. Natl. Acad. Sci. U.S.* **75**, 3118–3122
43. Yubisui, T. & Takeshita, M. (1980) *J. Biol. Chem.* **255**, 2454–2456
44. Kuwahara, S., Okada, Y., & Omura, T. (1978) *J. Biochem.* **83**, 1049–1059
45. Meldolesi, J., Corte, G., Pietrini, G., & Borgese, N. (1980) *J. Cell Biol.* **85**, 516–526
46. Okada, Y. & Omura, T. (1978) *J. Biochem.* **83**, 1039–1048
47. Borgese, N., Pietrini, G., & Meldolesi, J. (1980) *J. Cell Biol.* **86**, 38–45
48. Leskes, A., Siekevitz, P., & Palade, G.E. (1971) *J. Cell Biol.* **49**, 264–287
49. Orrenius, S. (1965) *J. Cell Biol.* **26**, 725–733
50. Sargent, J.R. & Vadlamudi, B.P. (1968) *Biochem. J.* **107**, 839–849
51. Omura, T. & Kuriyama, Y. (1971) *J. Biochem.* **69**, 651–658
52. Fujii-Kuriyama, Y., Negishi, M., Mikawa, R., & Tashiro, Y. (1979) *J. Cell Biol.* **81**, 510–519
53. Strittmatter, P., Rogers, M.J., & Spatz, L. (1972) *J. Biol. Chem.* **247**, 7188–7194

54. Enomoto, K. & Sato, R. (1973) *Biochem. Biophys. Res. Commun.* **51**, 1–7
55. Rogers, M.J. & Strittmatter, P. (1974) *J. Biol. Chem.* **249**, 5565–5569
56. Borgese, N. & Gaetani, S. (1980) *FEBS Lett.* **112**, 216–220
57. Okada, Y., Frey, A.B., Guenthner, T.M., Oesch, F., Sabatini, D.D., & Kreibich, G. (1981) *Eur. J. Biochem.* **122**, 393–402
58. Lowe, D. & Harrinan, T. (1973) *Biochem. J.* **136**, 825–828
59. Harano, T. & Omura, T. (1977) *J. Biochem.* **82**, 1541–1549
60. Gonzalez, F.J. & Kasper, C.B. (1980) *Biochemistry* **19**, 1790–1796
61. Harano, T. & Omura, T. (1978) *J. Biochem.* **84**, 213–223
62. Northemann, W., Schmelzer, E., & Heinrich, D.C. (1981) *Eur. J. Biochem.* **119**, 203–208
63. Rachubinski, R.A., Verma, D.P.S., & Bergeron, J.J.M. (1980) *J. Cell Biol.* **84**, 705–716
64. Bar-Nun, S., Kreibich, G., Adesnik, M., Alterman, L., Negishi, M., & Sabatini, D.D. (1980) *Proc. Natl. Acad. Sci. U.S.* **77**, 965–969
65. Gonzalez, F.J. & Kasper, C.B. (1980) *Biochem. Biophys. Res. Commun.* **93**, 1254–1258
66. Adelman, M.R., Blobel, G., & Sabatini, D.D. (1973) *J. Cell Biol.* **56**, 191–205
67. Ramsey, J.C. & Steele, W.J. (1976) *Biochemistry* **15**, 1704–1712
68. Adelman, M.R., Sabatini, D.D., & Blobel, G. (1973) *J. Cell Biol.* **56**, 206–229
69. Shibahara, S., Yoshida, T., & Kikuchi, G. (1979) *Arch. Biochem. Biophys.* **197**, 607–617
70. Chyn, T.L., Martonosi, A.N., Morimoto, T., & Sabatini, D.D. (1979) *Proc. Natl. Acad. Sci. U.S.* **76**, 1241–1245
71. Reithmeier, R.A.F., de Leon, S., & MacLennan, D.H. (1980) *J. Biol. Chem.* **255**, 11839–11846
72. Elder, J.H. & Morré, D.J. (1976) *J. Biol. Chem.* **251**, 5054–5068
73. Blobel, G. & Dobberstein, B. (1975) *J. Cell Biol.* **67**, 835–851
74. Blobel, G. & Dobberstein, B. (1975) *J. Cell Biol.* **67**, 852–862
75. Redman, C.M. & Sabatini, D.D. (1966) *Proc. Natl. Acad. Sci. U.S.* **36**, 608–615
76. Blobel, G. & Sabatini, D.D. (1970) *J. Cell Biol.* **45**, 146–157
77. Krieter, P.A. & Shires, T.K. (1980) *Biochem. Biophys. Res. Commun.* **94**, 606–611
78. Pickett, C.B., Rosenstein, N.R., Jeter, R.L., Morin, J., & Lu, A.Y.H. (1980) *Biochem. Biophys. Res. Commun.* **94**, 542–548
79. Kumar, A. & Padmanaban, G. (1980) *J. Biol. Chem.* **255**, 522–525
80. Bhat, K.S. & Padmanaban, G. (1979) *Arch. Biochem. Biophys.* **198**, 110–116
81. Lichtenstein, A.H. & Brecher, P. (1980) *J. Biol. Chem.* **255**, 9098–9104
82. Coleman, R. & Bell, R.M. (1978) *J. Cell Biol.* **76**, 245–253
83. Ito, A. & Sato, R. (1969) *J. Cell Biol.* **40**, 179–189
84. Omura, T., Noshiro, M., & Harada, N. (1980) in *Microsomes, Drug Oxidations, and Chemical Carcinogenesis* (Coon, M.J., Conney, A.H., Estabrook, R.W., Gelboin, H.V., Gillette, J.R., & O'Brien, P.J., eds.) Vol. 1, pp. 445–453, Academic Press, New York
85. Takesue, S. & Omura, T. (1970) *Biochem. Biophys. Res. Commun.* **40**, 396–401

86. Morgenstern, R., Meijer, J., DePierre, J.W., & Ernster, L. (1980) *Eur. J. Biochem.* **104**, 167–174

87. Heller, R.A. & Shrewsbury, M.A. (1976) *J. Biol. Chem.* **251**, 3815–3822

88. Prasad, M.R., Sreekrishna, K., & Joshi, V.C. (1980) *J. Biol. Chem.* **255**, 2583–2589

89. Moriyasu, M. & Ito, A. (1982) *J. Biochem.* in press

90. Boulan, E.R., Sabatini, D.D., Pereyra, B.N., & Kreibich, G. (1978) *J. Cell Biol.* **78**, 894–909

91. Akao, T. & Omura, T. (1972) *J. Biochem.* **72**, 1245–1256

92. Gold, G. & Widnell, C.C. (1976) *J. Biol. Chem.* **251**, 1035–1041

93. Ballas, L.M. & Arion, W.J. (1977) *J. Biol. Chem.* **252**, 8512–8518

94. Peterkofsky, B. & Assad, R. (1979) *J. Biol. Chem.* **254**, 4714–4720

95. Little, J.S., Thiers, D.R., & Widnell, C.C. (1976) *J. Biol. Chem.* **251**, 7821–7825

96. Cooper, M.B., Craft, J.A., Estall, M.R., & Rabin, B.R. (1980) *Biochem. J.* **190**, 737–746

97. Nilsson, O.S. & Dallner, G. (1977) *J. Cell Biol.* **72**, 568–583

98. Bollera, I.C. & Higgins, J.A. (1980) *Biochem. J.* **189**, 475–480

99. von Heijne, G. & Blomberg, C. (1979) *Eur. J. Biochem.* **97**, 175–181

100. von Heijne, G. (1981) *Eur. J. Biochem.* **116**, 419–422

101. von Heijne, G. (1981) *Eur. J. Biochem.* **120**, 275–280

102. Haugen, D.A., Armes, L.G., Yasunobu, K.T., & Coon, M.J. (1977) *Biochem. Biophys. Res. Commun.* **77**, 967–973

103. Botelho, L.H., Ryan, D.E., & Levin, W. (1979) *J. Biol. Chem.* **254**, 5635–5640

104. Koop, D.R., Persson, A.V., & Coon, M.J. (1981) *J. Biol. Chem.* **256**, 10704–10711

105. Black, C.D., French, J.S., Williams, Jr., C.H., & Coon, M.J. (1979) *Biochem. Biophys. Res. Commun.* **91**, 1528–1535

106. Du Bois, G.C., Appella, E., Armstrong, R., Levin, W., Lu, A.Y.H., & Jerina, D.M. (1979) *J. Biol. Chem.* **254**, 6240–6243

107. Gum, J.P. & Strobel, H.W. (1981) *J. Biol. Chem.* **256**, 7478–7486

108. Braell, W.A. & Lodish, H.F. (1981) *J. Biol. Chem.* **256**, 11337–11344

109. Ozols, J. & Gerard, C. (1977) *Proc. Natl. Acad. Sci. U.S.* **74**, 3725–3729

110. Mihara, K., Sato, R., Sakakibara, R., & Wada, H. (1978) *Biochemistry* **17**, 2829–2834

111. Roseman, M.A., Holloway, R.W., Calabro, M.A., & Thompson, T.E. (1977) *J. Biol. Chem.* **252**, 4842–4849

112. Enoch, H.G., Fleming, P.J., & Strittmatter, P. (1977) *J. Biol. Chem.* **252**, 5656–5660

113. Enoch, H.G., Fleming, P.J., & Strittmatter, P. (1979) *J. Biol. Chem.* **254**, 6483–6488

114. Rosbash, M. (1972) *J. Mol. Biol.* **65**, 413–422

115. Craig, R.K., Boulton, A.F., Harrison, O.S., Parker, D., & Campbell, P.N. (1979) *Biochem. J.* **181**, 737–756

116. Walter, P., Ibrahim, I., & Blobel, G. (1981) *J. Cell Biol.* **91**, 545–550

117. Walter, P. & Blobel, G. (1981) *J. Cell Biol.* **91**, 551–556

# Subject Index